T0341221

STEWARDING THE SOUND
The Challenge of Managing Sensitive Coastal Ecosystems

Editors

Leah Bendell
Professor
Biological Sciences, Simon Fraser University
Burnaby BC, Canada

Patricia Gallaugher
Adjunct Professor
Biological Sciences, Simon Fraser University
Burnaby BC, Canada

Shelley McKeachie
Past Chair/Director
Association of Denman Island Marine Stewards Society
Denman Island BC, Canada

Laurie Wood
Manager
Community Engagement and Research Initiatives
Faculty of Environment, Simon Fraser University
Burnaby BC, Canada

CRC Press
Taylor & Francis Group
Boca Raton London New York

CRC Press is an imprint of the
Taylor & Francis Group, an **informa** business

A SCIENCE PUBLISHERS BOOK

Cover credit: Cover illustration reproduced by kind courtesy of Prof. Leah Bendell (first editor)

CRC Press
Taylor & Francis Group
6000 Broken Sound Parkway NW, Suite 300
Boca Raton, FL 33487-2742

First issued in hardback 2019

© 2019 by Taylor & Francis Group, LLC
CRC Press is an imprint of Taylor & Francis Group, an Informa business

No claim to original U.S. Government works

ISBN-13: 978-0-367-11203-5 (hbk)

Visit the Taylor & Francis Web site at
http://www.taylorandfrancis.com

and the CRC Press Web site at
http://www.crcpress.com

Foreword

Isabelle M. Côté

Professor, Biological Sciences, Simon Fraser University, Burnaby, BC

It is a sight straight out of a David Attenborough documentary. The cobble beach is carpeted by the velvety sheen of herring spawn. Hundreds of gulls – glaucous-winged, Thayer's, Heerman's, mew – and oystercatchers are strutting, squawking, pecking at the seemingly endless supply of fish eggs. Just off the beach, a huge flotilla of monochrome surf scoters and outrageously ornate harlequin ducks take part in the smorgasbord of clams and cockles offered below the waves. The smooth, black heads of harbour seals and stellar sea lions bob at the surface and disappear. An orca exhales loudly as it breaks the surface of the water.

This is quintessential BC. Yet this superlative congregation of life is not found in a remote and inaccessible part of the coast. It occurs every spring in Baynes Sound – a short and shallow stretch of sea nestled between Vancouver Island and Denman Island in the Strait of Georgia. It would take just a couple of hours or less for a Vancouver city-dweller to drive there, and hundreds do every year to take in the natural spectacle.

Baynes Sound stands out as an extreme, on a provincial coast that ranks highly on the global wilderness and productivity scales. By some accident of geography and oceanography, the beaches of the Sound support an inordinate diversity and densities of bivalves. These, along with the spawning forage fish, attract globally significant numbers of waterbirds and wading birds, but also recreational and commercial interests. In fact, much of the Sound's beaches are under tenure for shellfish aquaculture, with plans for deep-water aquaculture currently under consideration. The area produces 39% of all oysters and 55% of all Manilla clams farmed in BC, and the industry employs some 600 workers – or about 10% of people who live nearby. According to a recent opinion survey by researchers at Vancouver Island University (D'Anna and Murray 2013), virtually all (95%) of Baynes Sound residents feel connected to nature, and while many recognize the benefits of aquaculture in terms of employment and food provision, more than half of agree that aquaculture has important impacts on beach ecology and most bemoan the lack of public access to beaches owing to shellfish growing operations.

In many ways, Baynes Sound epitomises the inherent difficulty people face everywhere when trying to simultaneously use and conserve natural resources. A balance needs to be achieved to satisfy the divergent interests of the multiple stakeholders in the area, e.g. residents, First Nations, the aquaculture business and a growing tourism industry, without impairing the key ecological processes that sustain this remarkable ecosystem. Unfortunately, there is increasing evidence that natural ecosystems have tipping points, that is they can withstand a range of human-made pressures relatively well, but beyond a certain cumulative threshold, they can collapse suddenly, without warning. The consequences of such a breakdown would be serious in ecological and economic terms, but also in the intangible terms of lost existence value for the residents of Baynes Sound and beyond. We have management tools at our disposal, such as rotational beach closures and the establishment of permanently closed areas, that can help us step back from the possible brink. Using them sooner is much, much better than later.

REFERENCES

D'Anna, L. and Murray, G. 2013. Baynes Sound opinion survey on shellfish aquaculture: Findings. Unpublished report, Institute of Coastal Research, Vancouver Island University. http://www2.viu.ca/icr/files/2012/06/Baynes-Sound-Opinion-Survey-on-Shellfish-Aquaculture-Report-20130927.pdf

Preface

DeFries and Nagendra (2017) first defined ecosystem management as a *wicked problem,* a problem that has no clear-cut solution. In trying to solve the wicked problem, two traps were identified; a tame solution and inaction from overwhelming complexity. The stewardship of Baynes Sound is a wicked problem with many complexities. We need to avoid the traps and take action immediately to protect and ensure the integrity of this marine ecosystem before it is too late.

REFERENCES

DeFries, R. and Nagendra, H. 2017. Ecosystem Management is a wicked problem. Science 356: 265-270.

Contents

Chapter 1

Introduction

L.I. Bendell[1], P. Gallaugher[2], S. McKeachie[3] and L. Wood[4]

[1]Professor, Biological Sciences, Simon Fraser University, Burnaby, BC
[2]Adjunct Professor, Biological Sciences, Simon Fraser University
[3]Past Chair/Director, Association of Denman Island Marine Stewards Society, Denman Island, BC
[4]Manager, Community Engagement and Research Initiatives, Faculty of Environment, Simon Fraser University

On coastal British Columbia (BC), between Denman Island and Vancouver Island, lies Baynes Sound, a rather understated name for one of coastal BC's most ecologically and biologically sensitive regions of the coast (Figure 1). It is a quiescent inland sea in contrast to the deep, drowned fjords that characterize BC's coast. It is comprised of estuaries, wetland complexes, expansive intertidal regions; features rarely seen along this coast. As a consequence of this habitat diversity, this region supports globally important populations of sea ducks and shore birds; forage fish including the pacific herring, a keystone species in the Pacific North East marine ecosystem; plus an epi and endo fauna community rich in a diversity of bivalves and other surface and subsurface intertidal species.

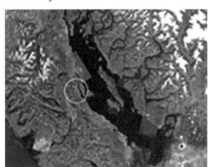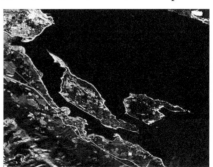

Figure 1 Baynes Sound and Lambert Channel.
(For color image of this figure, see Color Plate Section at the end of the book)

Corresponding Author: L.I. Bendell

It is also a region under increasing anthropogenic threats including an expanding shellfish industry, seaweed harvest and increased urbanization and industrialization that bring with them pollutants such as excess nutrients. Other stressors include the regulated and unregulated (poaching) wild shellfish harvest and increasing tourism as more people discover this part of BC's wild coast. Traditionally, we have looked to governments, both federal and provincial to manage and ensure sustainable resource use. However, lack of capacity has greatly hindered government's ability to manage this sensitive ecosystem. To address this gap, the Centre for Coastal Science and Management at Simon Fraser University hosted a workshop in April 2014. Its objective was to bring all stakeholders together with the hope of developing and mapping out a management approach that would protect Baynes Sound and provide a sustainable balance for all of the different interests. Although recommendations were made, none of them have been further developed or implemented (Bendell 2014). Again, in May of 2018, a similar workshop was held; this time organized by the Islands Trust and World Wildlife Fund Canada (WWF) with the same results (Hurley and Dunn 2018), as of this date.

Hence the collection of chapters presented herein. In an effort to stimulate action, presenters from the April 2014 meeting were invited to submit a chapter detailing their presentation. Contributors include representatives from the general public, academics, consultants, non government organizations (NGOs) and government scientists. Following this short introduction, Marjo Vierros and Lucie d'Amours describe three case studies of successful examples where communities come together and manage sensitive ecosystems in Japan and the Pacific Islands, and Canada's Magdalen Islands. Leah Bendell, Colin Levings and Ron Ydenberg then provide detailed descriptions of the uniqueness of the Sound. Threats to the Sound, primarily a result of economic pressures for resource development, are described by Ian Birtwell, Shelley McKeachie and Juan Alava. Duncan Knowler then shares an example of how economics can be a positive driver rather than a negative one in the use of sensitive ecosystems for economic gain. The co-editors conclude with a discussion on solutions, arising in part, from the open dialogue session where all stakeholders and participants from SFU's gathering expressed how a community may move forward to better manage and protect the Sound. It is our hope that this content may be applied to similar situations that others may be facing and through positive example demonstrate that positive change can be made.

ACKNOWLEDGEMENTS

The authors gratefully acknowledge funding from the Centre of Coastal Science and Management and the editorial assistance of the Biological Sciences office staff. We also thank Jerry Grey and Abigail Gossage of Ottawa for helping us develop the initial design of the book cover.

REFERENCES

Bendell, L.I. 2014. 'Stewarding the sound. Conveners' Report April 2014. It is the distance of a stone's throw. 39 pp. https://www.sfu.ca/coastal/research-series/listing/BaynesSoundSolutions.html

Hurley, K. and Dunn, K. 2018. Background information and a summary of the discussion that took place at the Baynes Sound/Lambert Channel Ecosystem Forum in Royston BC, 22-23 May 2018. 30 pp.

Chapter 2

Some Lessons Learned on Managing Multiple Stressors from Japan and the Pacific

Marjo Vierros

Director, Coastal Policy and Humanities Research, Vancouver, BC

INTRODUCTION

The management of sensitive ecosystems facing multiple stressors requires a holistic approach that takes into account the often complex linkages between ecosystems, species, physical and chemical processes, as well as human activities. Baynes Sound is home to unique and sensitive ecosystems and species, while also being subject to a multitude of human uses and development pressures. It is an example of an area where a holistic ecosystem approach is required to address cumulative environmental impacts. Because existing federal and provincial legislative and policy framework do not readily support application of an ecosystem approach, the communities living around Baynes Sound are left to explore options for putting in place local-level leadership in environmental management.

This paper discusses two case studies of community leadership in marine and coastal management that are rooted in cultural history and based on a long-term connection between a coastal area and its inhabitants. The first involves customary marine management approaches in the Pacific Island countries, while the second discusses the *satoumi* approach in Japan. These two case studies demonstrate how a holistic understanding of the environment, and an ecosystem approach to management can be undertaken by local communities to improve both biodiversity and livelihoods. This chapter examines the lessons that can be drawn from these case studies for the improved management of Baynes Sound.

The ecosystem approach in its modern-day format largely came about as a management response to a decline in biodiversity and natural resources, when approaches focusing on the management of single species had failed to produce the desired effects. Integrated coastal area management (ICAM) was one of the first contemporary ecosystem approaches specific to marine and coastal areas, and has been practiced worldwide for approximately fifty years (Vierros et al. 2006). Recent approaches, such as marine spatial planning, are now building on experiences with ICAM, and zoning of ocean space for various uses is taking place in the coastal areas and Exclusive Economic Zones of countries from Belgium to China. The Convention on Biological Diversity (CBD) and the Food and Agriculture Organization (FAO) have both adopted ecosystem approaches, one of them overarching, the other specific to the fisheries sector. Policy efforts to implement ecosystem approach are common under all multilateral agreements, although translation to practice in marine and coastal areas is still lagging behind terrestrial areas (Vierros et al. 2015).

While there are many different types of ecosystem approaches, as indicated above, it is generally recognized that there is no one "correct" way to implement an ecosystem approach, and that there is a great degree of flexibility in its application. Methods and tools depend on the specific problem or issue being addressed. However, all ecosystem approaches regardless of their origin seek to establish a balance between conservation and sustainable use, and acknowledge that humans and their cultural diversity are an integral part of the ecosystem. See, for example, decision V/6 of the Conference of the Parties to the Convention on Biological Diversity. In trying to understand how the ecosystem approach can be translated to practice, it is perhaps best to start with some of its earliest expressions: traditional management approaches.

INDIGENOUS WORLDVIEWS, MANAGEMENT SYSTEMS AND THE ECOSYSTEM APPROACH

While the ecosystem approach is still relatively new to contemporary resource management, where sectoral approaches and single species management were historically practised, it is intrinsic to most indigenous approaches. Many traditional societies have a holistic understanding of the coastal area. In this worldview, the land, sea, freshwater and atmosphere, as well as humans and their cultural and spiritual values, are interconnected and part of one system.

For example, some indigenous peoples in Australia view the coastal landscape as an integrated cultural landscape/seascape where no distinction is made between land and sea. Barber (2005) provides the following description of Blue Mud Bay in the Northern Territory: "In Blue Mud Bay, much of daily life and activities occur in the context of the flow of water, from freshwater rivers which flow into the increasingly salty water of the sea, and the seasonal cycles of rain and storms. These, in turn, affect the life cycles of species both on sea and land that provide food for the Aboriginal communities of Blue Mud Bay. The environments of the land and sea, their seasonality, flows and the animal and human communities that they support are all interrelated, and viewed in a holistic manner by the inhabitants of the area."

A worldview where all things are interconnected has resulted in management systems that are holistic, and are expressions of what we today call the ecosystem approach. One example of this is the native Hawaiian concept of *ahupua'a*. *Ahupua'a* are roughly wedge-shaped catchments extending from upland tropical forests to the fringing coral reefs and nearshore ecosystems (Kaneshiro et al. 2005). The *ahupua'a* contained nearly everything Hawaiians required for survival. Within the *ahupua'a*, a wise conservation system was practiced by the caretakers to prevent exploitation of the land and sea while allowing the people to use what they needed for sustenance. Adaptive management practices were employed, keyed to subtle changes in natural resources, and sophisticated social controls on resource utilization were an important component of the system. Stewardship of resources was supported by a sense of responsibility, translating elements of Hawaiian spirituality into the natural landscape. Amid a belief system that emphasized the interrelationship of elements and beings, the *ahupua'a* contained those interrelationships in the activities of daily and seasonal life.

Counterparts to *ahupua'a* are found in other mountainous Pacific Islands, including *vanua* in Fiji and *tapere* in Cook Islands. The *ahupua'a* system is depicted in Figure 1.

This holistic worldview, and a deep understanding of the environment and ecology of a place, gave rise to traditional management approaches that are often still implemented by communities in Pacific Island countries.

TRADITIONAL MANAGEMENT SYSTEMS IN THE PACIFIC ISLANDS

In the Pacific Islands, traditional management systems have been implemented for thousands of years, based on traditional spiritual beliefs and established by traditional leaders. These management approaches consist of a multitude of tools such as temporary closed areas, seasonal bans on harvesting, species/catch restrictions, watershed management, traditional agriculture systems, cyclone preparation, and other similar approaches. Closed areas include the *tabu* areas of Fiji, Vanuatu and Kiribati, the *kapu* in Hawaii, the *ra'ui* in Cook Islands, the *tambu* in Papua New Guinea, the *mo* in the Marshall Islands, the *fono* in Niue and the *bul* in Palau (Vierros et al. 2010; Govan et al. 2009; Parks and Salafsky 2001). The application of traditional management systems is based on intimate, long-term knowledge of local species and ecosystems, and their ecological relationships. Importantly, they are flexible and can be applied quickly in response to changing environmental conditions, making them a form of adaptive management.

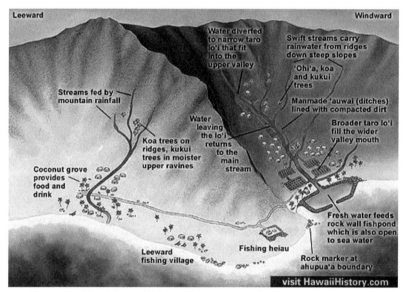

Figure 1   Depiction of *ahupua'a* as a watershed management unit
(From HawaiiHistory.com).

While traditional knowledge and associated management systems in the Pacific Islands deteriorated, and in some cases lost, during the colonial period, there has been a recent revival (Govan et al. 2009; Vierros et al. 2010). These revitalized traditional practices have evolved through the years in response to societal and economic changes (Johannes and Hickey 2004). Like most places, Pacific Islands have recently experienced environmental degradation, habitat loss, land-based pollution, fisheries decline, climate change impacts, population growth and food security concerns. The revival of traditional management systems has been to a great degree a response to these increasing impacts, which are perceived by the communities to achieve livelihoods benefits (Govan et al. 2009). This response is highly significant, leading to a proliferation of marine managed areas implemented by communities in the Pacific Islands. Their extent is difficult to track because many are not reported to or by governments, or included in international databases, but a 2009 survey found that they spanned 15 independent Pacific Island states and 500 communities, covering approximately 30,000 km^2 of area (Govan et al. 2009). These areas are actively managed by communities and resource-owning groups, or collaboratively by resident communities with local government and/or partner organizations. They are increasingly being recognized in international processes, for example the 2017 United Nations Ocean Conference voluntary commitments included 15 that were specific to areas managed by local communities.

The primary function of traditional management systems was not conservation for its own sake, but instead management was intended to maintain and improve livelihoods, food security and local economic revenue. This is still the case today. Conservation and sustainable use are considered interlinked and inseparable by communities in the Pacific Islands, and are components of the surviving concepts of traditional environmental stewardship (Govan et al. 2009). For example, temporary closed areas, such as *tabu* areas, were generally harvested after the resource had recovered. Additionally, they provided food for special events or a cache for when resources on regular fishing grounds ran low. While the main goal of community managed marine areas today is to benefit communities, they have in most cases also been successful in delivering biodiversity and fisheries outcomes (Govan et al. 2009; Vierros et al. 2010). In Fiji, extensive monitoring has demonstrated both the economic and fisheries impacts, with 20-fold increases in clam density in *tabu* areas since 1997, tripling of fish catches, and 200-300% increase in harvest in adjacent areas. In the same time period, household income has seen a 35-45% increase (Aalbersberg et al. 2005).

The spread and endurance of these locally managed marine areas can be attributed in great part to the perception by communities that benefits are, or will likely be, achieved, thus offsetting the initial cost and effort of their establishment. Their inception is based on a community awareness of a need for action, ensuring a high degree of support locally. In a global sense, they provide an alternative to the top-down, government-led approach to conservation, which has in many cases led to the establishment of "paper parks". They are more cost effective to implement for Pacific Island countries, with their long coastlines and minimal budgets for environmental management, and are self-sustaining through community leadership (Govan et al. 2009).

For British Columbia, the lessons that can be drawn from the Pacific Islands experience include the potential effectiveness of bottom-up approaches, their replicability both within similar cultures and ecosystems and beyond (Rocliffe et al. 2014 on LMMAs in the Western Indian Ocean), their cost-effectiveness, their ability to find a balance between conservation and sustainable use, and the potential they have to provide benefits for local livelihoods, resources and biodiversity. The Pacific Island approaches are rooted in a long history of traditional knowledge and management, and communities in most cases have ownership of and/or tenure rights to marine areas. In British Columbia, control over marine and coastal areas rests with the federal and provincial governments, thus limiting the options communities have to act on their own. Municipal governments and community organizations have some avenues to control land-based activities impacting on adjacent marine areas. Traditional knowledge and practices exist within the First Nations communities on the coast, and they should be seen as important partners in management activities.

SATOUMI IN JAPAN

Unlike the Pacific Islands, Japan is a developed country with high population density and a highly altered coastline. Japan's seas have suffered from multiple human impacts that include fisheries decline, pollution and loss of coastal habitat to seawalls and other development (UNU-IAS 2011). In response, the Japanese government, scientists, fishermen and other community members have supported the practice of *satoumi*, which is an ecosystem approach to coastal management based on Japanese cultural heritage and traditional ecological knowledge.

Like the traditional Pacific Islands approaches described previously, *satoumi* is centered on providing benefits to both people and biodiversity. In Japanese, "Sato" means the area where people live, while "Umi" means the sea. Human interaction with the coastal environment nurtures nature and contributes to the enhancement of biodiversity. When *satoumi* is restored in coastal waters, marine productivity and biodiversity are enhanced through the involvement of, and in harmony with, people (Yanagi 2008). Achievement of *satoumi* relies on a long cultural heritage of fishers' knowledge and management, and an understanding of the interactions within and between ecosystems and human communities in the coastal zone. Thus, the achievement of *satoumi* relies on human stewardship in a similar way that management of *ahupu'a* and other traditional Pacific Island management systems do.

The Figure 2 describing *satoumi* is very similar to that depicting the Hawaiian *ahupua'a* above. The only difference is that Japan's coastline is highly developed with a large amount of urban area.

While traditional management systems in the Pacific Islands were generally bottom-up systems implemented by village chiefs, *satoumi* contains a mixture of both top-down and bottom-up elements. Unlike many other management practices based on traditional cultural heritage, *satoumi* been incorporated into Japanese national policies, including the Strategy for an Environmental Nation in the 21st Century (2007), the

Third National Biodiversity Strategy of Japan (2007), and the Basic Plan on Ocean Policy (2008). The concept is flexible and inclusive enough to be used as a basis for mainstreaming, and thus different government departments, including the Ministry of Land, Infrastructure, Transport and Tourism (MLIT); the Fisheries Agency (FA); and the Ministry of the Environment (MOE) each define and implement their own versions of *satoumi* and get together regularly through a coordinating committee to discuss their *satoumi* activities (Ota et al. 2011).

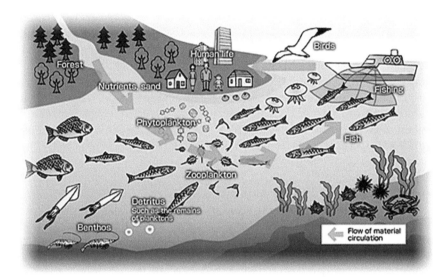

Figure 2   Depiction of *satoumi*, including connections between land and sea (Ministry of Environment, Japan).

Because the term "*satoumi*" instantly evokes a certain romantic ideal for Japanese people (UNU-IAS 2011), it is also a way of "branding" certain types of management activities in a way that is both recognizable and memorable. *Satoumi* complements restriction-based conservation through a variety of activities undertaken in an ecosystem approach context. *Satoumi* activities include planting eelgrass to restore coastal ecosystems, re-planting forests in watershed areas, sustainable cultivation of oysters, public education, and working with fishing communities to revive traditional fishing methods. Environmental education is central through participation of community and schools in restoration and monitoring activities. Many satoumi projects have applied zoning schemes and/or marine spatial planning (UNU-IAS 2011).

While *satoumi* is part of government policies, and provides modest government funding for environmental restoration and monitoring activities, it also encourages community self-initiative. For example, local fishers are seen as integral components of ecosystems with an active role to play in ecosystem-based management. Fishermen understand the connection between the mountains and the sea, the inshore and offshore better than most coastal inhabitants. The ancient 17th century tradition of *uotsukirin*, or fish-breeding forest demonstrates this understanding. Though already widely practiced in the 17th century, *Uotsukirin* was formalised into the Japanese legal system during the Meiji Era in 1897 as part of the forest law, and provides the basis for modern-day watershed and forest restoration activities undertaken by fishermen (Matsuda 2011). The expression, *"Mori wa umi no koibito"* (the forest is the sweetheart of the sea) (Hatakeyama 1994) describes the intimate connection between mountains, rivers and coastal ecosystems, and has become a kind of catchphrase reminding fishers to turn their eyes to the mountains (Tsujimoto 2011). Accordingly, fishers' cooperatives undertake habitat restoration activities, such as re-planting forests in watersheds to manage run-off, and planting seagrass and *Sargassum* weed to restore coastal spawning and nursery habitat. In addition, fishermen implement voluntary no-take areas to manage coastal fisheries. These activities show a deep understanding of the need to share the costs and benefits of caring for a resource, which is the similar to the notion of stewardship seen in the Pacific Island countries.

In addition to habitat restoration activities undertaken by fishermen and other local communities, ongoing environmental monitoring is carried out by a partnership of citizens, scientists and local governments. Targeted scientific research is an important component of *satoumi*, and scientists have been instrumental in carrying out the research necessary for prioritizing activities in support of improved water quality and biodiversity, and in monitoring the results. Local knowledge, and in particular fishermen's knowledge, is used in addition to science. Scientists have made a long-term commitment to assisting local communities, thus providing them with the support necessary for management activities.

Mechanisms for implementing local *satoumi* projects include coordinating bodies, which have a broad range of stakeholders as members. Compliance mechanisms are based on peer monitoring and sanctions imposed by community stakeholders (Yagi et al. 2010; Yagi 2011). The peer monitoring and compliance system works because the same group of stakeholders bears both the costs of a management action and enjoys its benefits in the long term. Peer monitoring and enforcement are also made possible in the context of fisheries no-take areas due to historical territorial use rights granted to fisheries cooperatives through legislation (Yagi 2011). In general, users have an interest in the long-term sustainability of a resource on an intergenerational timescale, and thus their investment today will result in benefits to generations to come.

Satoumi is an important example for British Columbia because it is implemented in an area with high population density, and a multitude of uses of the coastal zone. British Columbia lacks the historical sea tenure arrangements, which have enabled Japanese fishermen to undertake management activities with a view towards a long-term stewardship of the resource, including relying on voluntary closed areas, habitat restoration, peer monitoring and enforcement. Many of British Columbia's industries are likely too new to have afforded their operators full understanding of how their activities impact environment and ecology, and assessment of cumulative impacts is also lacking. Similarly, a sense of ownership of place, and the associated need to act as its stewards, may take considerable time to develop, and may not be possible in cases where a short-term profit motive by companies or entities based in distant locations drives development. Regardless, a regime where resource users share in the costs of management should be achievable in British Columbia. The *satoumi* experience also shows the importance of having local communities and other stakeholders at the forefront of environmental management and decision-making.

Other lessons learned from *satoumi* projects include the need for inclusivity in management through incorporation of all community stakeholders, scientists and government entities, including building long-term cooperation and partnerships towards management. The flexibility and "branding" of the *satoumi* concept have also helped bring people together. Similar branding that relies on pride in environment, culture and history may also be possible on the British Columbia coast.

CONCLUSIONS

The overarching question is whether, and to what degree, the approaches described here can have relevance to British Columbia, and particularly to the management of Baynes Sound. There is generally no one-size-fits-all solution to implementing an ecosystem approach in management. The right approach, or combination of approaches, depends on local circumstances that may include cultural history, resource dependence, environmental pressures, ownership structures and governmental policies.

There are, however, commonalities between cultures and marine management and stewardship strategies across the Asia-Pacific. The approaches taken in the Pacific Islands and in Japan are similar in their reliance on community-based approaches, a holistic vision that extends from the mountain to the sea, use of local/traditional ecological knowledge along with scientific knowledge, and an ethic of stewardship that is based on a notion of intergenerational equity. These approaches have often been effective in producing social, economic and biodiversity benefits. Traditional ecological knowledge and practices, as well as a sense of stewardship, also exist among the First Nations on the British Columbia coast (e.g. Jones et al. 2010; Turner et al. 2000), as does a desire to protect and manage resources that communities have traditionally depended on (Jones et al. 2017). Much could be learned from these traditional approaches in moving towards long-term sustainable management of the British Columbia coast. Gaining a better understanding of cultural connections across the Asia-Pacific, as well as of the accumulated knowledge and wisdom locally, can foster a sense of affinity and identity, and could stimulate stakeholder involvement in marine management.

Some of the general lessons learned from the Pacific Islands and Japan that may be transferable to Baynes Sound include:

Management is often best undertaken at the local level. People generally develop a strong connection with the place where they live, and this is particularly true of places that have been inhabited by the same group through generations. Communities depend on the resources of the surrounding land and sea, understand the area's environment and ecology, and its cultural and spiritual values better than anyone else. Their stake in the area and its environmental health and resilience is direct and immediate.

Enabling policies and programmes by government agencies can greatly support and empower local management. This was, in particular, demonstrated by the *satoumi* example in Japan. In British Columbia, federal and provincial government leadership on environmental issues is currently lacking, and thus local management may need to rely on support from municipal government and civil society groups, while finding ways to make higher governmental levels more responsive to their needs, and to link local efforts to broader developments provincially, nationally and internationally.

Successful management efforts generally have coordinating bodies that bring together all stakeholders and rights-holders in an area. They are inclusive and seek to incorporate a diversity of views, including industry, conservation, local community, First Nations and scientists. The goal of these bodies is to build partnerships towards long-term priorities that may include community and ecosystem resilience, livelihoods, and adaptation to climate change.

A holistic view encompassing the environment, species and people and their interactions is necessary to address all impacts and their cumulative effects on an area. An ecosystem approach to management, rather than a single species approach, is needed to deal with complex problems, such as those impacting Baynes Sound. This approach was demonstrated in both the Pacific Islands and Japanese examples. The traditional Pacific Islands approaches demonstrated that knowledge, beliefs and practices that contributed to resource management pervaded all facets of life, including spiritual, cultural and social systems, where all things and events were inherently connected (Hickey 2006).

A commitment to long-term collaboration between scientists and communities is important to ensure that management is based on best available science. Close interaction between scientists and community members also ensures that science is directed, relevant, timely and easily accessible. The Pacific Islands and Japanese examples demonstrated the importance of traditional and local knowledge in management, in addition to scientific knowledge.

Intergenerational equity should be a guiding principle of management. Management can only be successful if it has as its goal long-term sustainability rather than short-term profit. The concept of stewardship that exists traditionally in cultures in the Pacific Islands, Japan and on the British Columbia coast could be extended to a responsibility of resource users to contribute to the sustainable management of the resource.

REFERENCES

Aalbersberg, B., Tawake, A. and Parras, T. 2005. Village by village: Recovering Fiji's coastal fisheries. pp. 144-152. *In*: World Resources 2005: The Wealth of the Poor: Managing Ecosystems to Fight Poverty. United Nations Development Programme, United Nations Environment Programme, The World Bank, World Resources Institute, Washington, DC.

Barber, M. 2005. Where the clouds stand: Australian aboriginal relationships to water, place, and the marine environment in blue mud bay, Northern Territory.

Govan, H., Tawake, A., Tabunakawai, K., Jenkins, A., Lasgorceix, A., Schwarz, A.M., Aalbersberg, B., Manele, B., Vieux, C., Notere, D. and Afzal, D.. 2009. Status and potential of locally-managed marine areas in the South Pacific: Meeting nature conservation and sustainable livelihood targets through wide-spread implementation of LMMAs. SPREP/WWF/WorldFish-Reefbase/ CRISP. 95 pp + 5 annexes.

Hatakeyama, S. 1994. The Forest is the Sweetheart of the Sea. Tokyo: Hokuto Shuppan (In Japanese)

Hickey, F.R. 2006. Traditional marine resource management in vanuatu: Acknowledging, supporting and strengthening indigenous management systems. SPC Traditional Marine Resource Management and Knowledge Information Bulletin, No. 20, pp. 11-23. http://www.spc.int/coastfish/News/ Trad/20/Trad20_11_Hickey.pdf

Johannes, R.E. and Hickey, F.R. 2004. The evolution of village-based marine resource management in Vanuatu between 1993 and 2001. Coastal region and small island papers 15. UNESCO, Paris, France. http://www.unesco.org/csi/ wise/indigenous/vanuatu1.htm

Jones, R., Rigg, C. and Lee, L. 2010. Haida marine planning: first nations as a partner in marine conservation. Ecology and Society 15(1): 12. http://www.ecologyandsociety.org/vol15/iss1/art12/

Jones, R., Rigg, C. and Pinkerton, E. 2017. Strategies for assertion of conservation and local management rights: A Haida Gwaii herring story. Marine Policy 80: 154-167. https://doi.org/10.1016/j.marpol.2016.09.031

Kaneshiro, K.Y., Chinn, P., Duin, K.N., Hood, A.P., Maly, K. and Wilcox, B. 2005. Hawai'i's mountain-to-sea ecosystems: social–ecological microcosms for sustainability science and practice. EcoHealth 2(4): 349-360.

Matsuda, O. 2011. Uotsukirin—the fish-breeding forest. p 36. *In*: United Nations University Institute of Advanced Studies Operating Unit Ishikawa/Kanazawa. 2011. Biological and Cultural Diversity in Coastal Communities, Exploring the Potential of Satoumi for Implementing the Ecosystem Approach in the Japanese Archipelago. Secretariat of the Convention on Biological Diversity, Montreal, Technical Series no. 61.

Ota, Y., Chiba, Y. and Dolan, J. 2011. Mainstreaming *satoumi* in Japanese national policy: Introduction to the case studies. pp 8-14. *In*: United Nations University Institute of Advanced Studies Operating Unit Ishikawa/Kanazawa. 2011. Biological and Cultural Diversity in Coastal Communities, Exploring the Potential of Satoumi for Implementing the Ecosystem Approach in the Japanese Archipelago. Secretariat of the Convention on Biological Diversity, Montreal, Technical Series no. 61.

Parks, J.E. and Salafsky, N. 2001. Fish for the future? A collaborative test of locally-managed marine areas as a biodiversity conservation and fisheries management tool in the Indo-Pacific region: Report on the initiation of a learning portfolio. The World Resources Institute. Washington DC.

Rocliffe, S., Peabody, S., Samoilys, M. and Hawkins, J.P. 2014. Towards a network of locally managed marine areas (LMMAs) in the Western Indian Ocean. PloS one, 9(7): p.e103000

Tsujimoto, R. 2011. Fisher activities to conserve the ecosystem of Toyama Bay. pp. 30-37. *In*: United Nations University Institute of Advanced Studies Operating Unit Ishikawa/Kanazawa 2011. Biological and Cultural Diversity in Coastal Communities, Exploring the Potential of Satoumi for Implementing the Ecosystem Approach in the Japanese Archipelago. Secretariat of the Convention on Biological Diversity, Montreal, Technical Series no. 61.

Turner, N.J., Ignace, M.B. and Ignace, R. 2000. Traditional ecological knowledge and wisdom of aboriginal peoples in British Columbia. Ecological Applications 10: 1275-1287.

United Nations University Institute of Advanced Studies (UNU-IAS) Operating Unit Ishikawa/Kanazawa. 2011. Biological and Cultural Diversity in Coastal Communities, Exploring the Potential of Satoumi for Implementing the Ecosystem Approach in the Japanese Archipelago. Secretariat of the Convention on Biological Diversity, Montreal, Technical Series no. 61. http://www.cbd.int/doc/publications/cbd-ts-61-en.pdf

Vierros, M., Douvere, F. and Arico, S. 2006. Implementing the ecosystem approach in open ocean and deep sea environments. A United Nations University – Institute of Advanced Studies Report. http://unu.edu/publications/policy-briefs/implementing-the-ecosystem-approach-in-open-ocean-and-deep-sea-environments.html

Vierros, M., Tawake, A., Hickey, F., Tiraa, A. and Noa, R. 2010. Traditional marine management areas of the pacific in the context of national and international law and policy. United Nations University—Traditional Knowledge Initiative, Darwin, Australia.

Vierros, M., Cresswell, I.D., Bridgewater, P. and Smith, A.D.M. 2015. Ecosystem approach and ocean management. pp. 127-145. *In*: Aricò, S. (ed.). Ocean Sustainability in the 21st Century. Cambridge University Press. DOI: 10.1017/CBO9781316164624.009

Yagi, N, Takagi, T., Takada, Y. and Kurokura, H. 2010. Marine protected areas in Japan: Institutional background and management framework. Marine Policy 34: 1300-06.

Yagi, N. 2011. *Satoumi* and institutional characteristics of Japanese coastal fishery management. pp 96-101. *In*: United Nations University Institute of Advanced Studies Operating Unit Ishikawa/Kanazawa (2011). Biological and Cultural Diversity in Coastal Communities, Exploring the Potential of Satoumi for Implementing the Ecosystem Approach in the Japanese Archipelago. Secretariat of the Convention on Biological Diversity, Montreal, Technical Series no. 61.

Yanagi, T. 2008. Sato-Umi– A new concept for sustainable fisheries. pp. 351-358. *In*: Tsukamoto, K., Kawamura, T., Takeuchi, T., Beard, T.D., Jr. and Kaiser, M.J. (eds). Fisheries for Global Welfare and Environment. 5th World Fisheries Congress, 2008. Terrapub.

Chapter 3

ZIP Committee of Magdalen Islands: Integrated management of our coasts and sensitive inland water bodies

Lucie d'Amours

Past Chair, Magdalen Islands ZIP Committee,
Retired biology professor and Academic advisor,
Cégep de la Gaspésie et des Îles, Québec.

THE AREA

The Magdalen Islands (Îles de la Madeleine) form a small archipelago in the Gulf of St. Lawrence with a land area of 205.53 km^2 (79.36 mi^2), of which 60 are sand dunes. If we include inland water bodies and shallow tidal zones, the area covers 360 km^2. Though closer to Prince Edward Island (Île-du-Prince-Édouard) and Nova Scotia (Nouvelle-Écosse), the Islands are in the Canadian province of Quebec as seen in Figure 1. This archipelago includes about fifteen islands, seven of which are "joined" by sand dunes that include water bodies called "lagunas" or "basins" (Remillard et al. 2016).

A few small islands are isolated from the main "central knot" (Figure 1). One of them, Entry Island (Île d'Entrée), is still inhabited by an anglophone community from Irish and Scottish origin. Brion Island and Bird Rock are also found in these isolated islands, the first one being an ecological reserve, and the second one a protected migratory bird area.

Together the islands represent a region of tremendous habitat and species diversity. Unlike the rest of the Gulf of St. Lawrence, which is currently undergoing an enhancement, the Islands are sinking, causing a rise in sea level greater than that related to climate change (about 3.2 mm/year) (NASA 2018). Adding to this change is the rate of erosion of the sandstone cliffs. This erosion is being exacerbated by the lack of protection from ice due to milder winters in recent years.

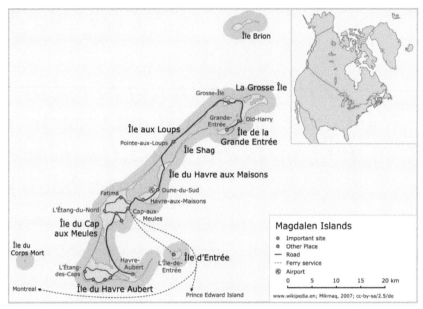

Figure 1   Magdalen Islands. Image from Wikepedia. Accessed November 2018. *(For color image of this figure, see Color Plate Section at the end of the book)*

The area is extensively monitored by the scientific community with approximately 450 km of diverse coastline encircling the main islands of the archipelago and making it a "lab area" of great interest. Scientists estimate that the archipelago will lose 80 meters of sand dunes and 38 meters of cliffs by 2050 (Friesinger and Bernatchez 2010).

Each of the islands has a rocky core, usually raised in the center with small hills, the highest point being Big Hill on Entry Island, at 174 meters. Salt domes support the islands and are believed to shelter significant amounts of gas and oil. Electricity, on the islands, comes from thermal power relying on oil imports. Efforts are being deployed to develop more renewable energy sources. For example, the wind is omnipresent in the Islands and while it may be a limiting factor for many species, it presents great opportunity for wind power. Energy from biomass, natural gas, submarine cable from the mainland and waste diversion are also being analyzed (Municipalité des Îles-de-la-Madeleine 2018).

The Islands also have an abundance of high-quality freshwater. It is the only area in Quebec, where the community fully draws its drinking water from groundwater where freshwater forms a fragile hydrostatic equilibrium on top of saltwater.

The Islands are home to many kinds of terrestrial and aquatic ecosystems and provide ecological services such as spawning habitat for fish, mollusks, and crustaceans; support of primary productivity through extensive eelgrass beds, berry crops, nesting areas, endangered species such as the piping plover, marine mammal rest areas, and a unique food web.

THE COMMUNITY

The actual living area in the archipelago is small with 160 km^2 located on the rocky core and supporting approximately 13,000 people and 60,000 tourists annually. The small size of this area necessitates linkages to the rest of the continent for both importation of goods and exportation of local harvests. The Islands enjoy regular boat transportation from Prince Edward Island and also cities from the Laurentian corridor such as Quebec and Montreal. A small airport also permits an impressive daily traffic of goods and people.

The fishing industry is the foundation of the economy as it generates revenues of nearly $115 million and employs more than 2,100 people. Almost a third of the workforce of the Magdalen Islands works in the commercial fishing industry and fish and seafood processing. The fisheries are mainly lobster and crab. There are approximately 400 fishing companies flourishing, mostly for lobster fishing. Aquaculture businesses have ten business licenses and are active primarily with mussels, scallop and oysters. Seventy per cent of the value of landings in lobster and mariculture for all Quebec comes from the archipelago. The other pillar of the economy is the tourist industry employing roughly 1,750 people, mainly during summer. This industry is growing each year as more international cruises dock, extending the tourists season (Municipalité des Îles-de-la-Madeleine 2018).

PROTECTING THE FRAGILE ISLAND ENVIRONMENT

The fisheries and tourism economies revolve around the ubiquitous natural resources in the area: strengthening the economy must therefore be consistent with the preservation of the natural environment. Since 1980, many organizations are working towards this goal. Some are dedicated to recycling waste products, some to preserving habitats and species and others to educating people. One conservation society buys land for long-term protection, while a local research center (CERMIM —Centre de recherche sur les milieux insulaires et maritimes), helps in linking the community to scientists in many different fields. Another centre specializes in innovation in aquaculture and fisheries (MERINOV —Centre d'Innovation de l'aquaculture et des pêches du Québec).

In general, island communities are used to responding and adapting to challenges because often problems arise more quickly, sometimes have a greater impact, and are more visible on islands than on the mainland.

In contrast to this advantage, many government managers responsible for the public lands of Magdalen Islands (30% of its area) are more familiar with mainland resources such as forest management than they are with island sand dune management especially given that their head offices are on the mainland of Quebec or other provinces. To address this, agreements were signed for management delegation to the municipal council of the Islands recognizing the local expertise was better able to account for the peculiarities of the region as well as to recognize the maritime community and its insularity context.

Because of the isolation of the Magdalen Islands, Islanders acknowledge the need to be self-sustaining in as many ways as possible. The ZIP mandate is to promote knowledge of the St. Lawrence waterway including its river, gulf and coastal environments, and encourage community involvement.

THE ZIP COMMITTEE AND INTEGRATED MANAGEMENT OF INLAND WATER BODIES

Along the St. Lawrence River, thirteen ZIP (Zone d'Intervention Prioritaire) Committees were formed in zones identified as being of priority importance (Figure 2). These groups, comprised of local residents, work in collaboration and have a mandate to promote knowledge about the St. Lawrence River and to encourage local initiatives of protection, restoration, conservation and enhancement of the uses and resources of the St. Lawrence that comply with sustainable development. Their work is part of the St. Lawrence Action Plan.

In 1988, the governments of Canada and Quebec began working with various stakeholders "to conserve, restore, protect and develop the St. Lawrence River." (St. Lawrence action plan 2011-2016). This agreement is now in its fifth phase, bringing together different federal and provincial departments with mandates to improve water quality, biodiversity and control uses of the St. Lawrence River.

The Magdalen Islands ZIP Committee structure is comprised of fourteen members with representatives from the public, health, education, recreation, municipal government, environment, and industry sectors. The committee was charged with developing an action plan for the islands (Ecological Rehabilitation Action Plan, ERAP) which identified eighteen issues, one of which was to establish an integrated management plan for the inland bodies of water and to alert users of these waters to necessary actions.

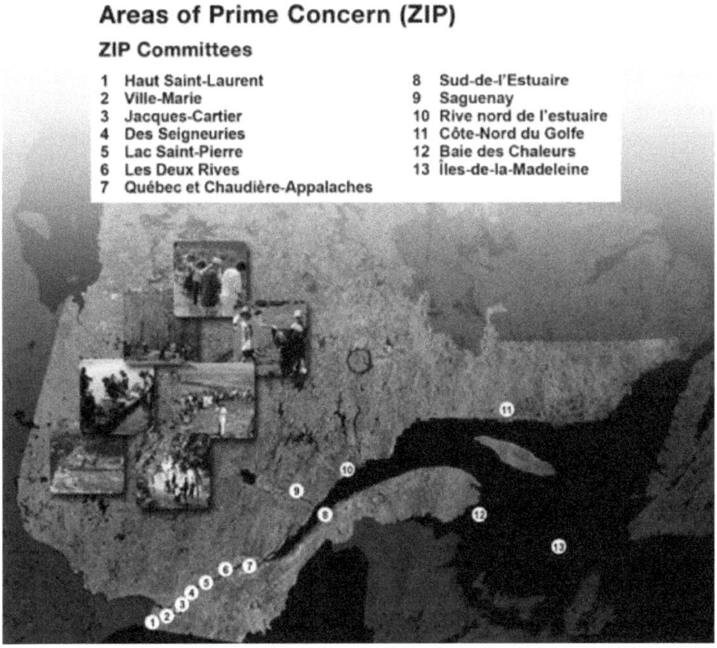

Areas of Prime Concern (ZIP)

ZIP Committees

1 Haut Saint-Laurent	8 Sud-de-l'Estuaire
2 Ville-Marie	9 Saguenay
3 Jacques-Cartier	10 Rive nord de l'estuaire
4 Des Seigneuries	11 Côte-Nord du Golfe
5 Lac Saint-Pierre	12 Baie des Chaleurs
6 Les Deux Rives	13 Îles-de-la-Madeleine
7 Québec et Chaudière-Appalaches	

Figure 2 ZIP committee locations (St-Lawrence action plan 2011-2016). *Courtesy*: http://planstlaurent.qc.ca/en/integrated_management/zip_program.html

Five separate integrated management committees representing each body of water were formed and brought stakeholders together. Repeated interaction on the committee led to the development of trust and respect for various users and fostered collaboration in the identification of common priorities of all users (Bergeron 2007). The idea behind the integrated management process is that users, members of the ZIP Committee and managers (three levels, federal, provincial and municipal) have a neutral place to exchange information and to develop sustainable practices in natural resource management.

Their initial project was to "understand the past to manage the future" in an effort to sustainably manage each body of water. The first task in reaching this goal was to make an inventory of the ecological and socio-economic aspects of the five bodies of water. Using GIS, this involved the mapping of the natural (i.e., flora, fauna), physical, and human environment. This work led to the conservation of our inland water bodies by improving current knowledge and involving users and stakeholders in an integrated management approach.

Methodology and Example (Bergeron 2007)

The meetings of these committees were conducted by the ZIP Committee to ensure a neutral and collaborative space where all members had an equal voice at the table. The meetings took place face-to-face, by conference call, email, local venues or whatever way worked best. They were often held on Sunday (a day when fishermen are not usually on the sea), and in a festive location to encourage participation (a local brewery, for example). The first meetings encouraged the exchange of information or requests for more information on a given subject to create a level playing field of knowledge. For instance, the follow-up of the work done by federal (Department of Fisheries and Oceans) and provincial (MAPAQ) department of fisheries and aquaculture, on the carrying capacity of the water bodies was very helpful in preventing misinformation and misunderstandings. (This process also led to the drafting of a code of good practices for fish farmers.)

Then, based on the discussions, a list of actions was drafted for each body of water. Each action was then examined outside of the meetings to be clarified and classified as being unique or common to a specific zone. For example, *cleaning of the embankments of the water body* was a common recommendation to all bodies of water.

The various actions expressed in the five integrated management plans and in the ERAP are united under four main themes:

1. Human activities and water quality;
2. Fauna and flora;
3. Land protection, conservation and enhancement;
4. Awareness and communication.

For each action identified, the members had to qualify the actions as:

- very important (3),
- important (2),
- not so important (1) or
- not important (0)

on the environmental, social and economic viewpoint. The ZIP Committee then settled priority on the actions with these criteria:

- importance of the action for the members (9 points);
- total, partial or zero achievement of the action (5 points);
- number of water bodies concerned by the action (only for the common actions) (5 points);
- correspondence with a priority from ERAP (1 point).

An integrated action plan was then made for each committee and presented to them for validation. It is hoped that this process creates balanced solutions minimizing conflicts of use and leading to long-term use of natural resources, with all partners on board, and all the energy focussed in a positive process that respects local ways of doing things. For the managers, this is a powerful orienting and decision-making tool, that ensures the support of the local community. This also links local priorities with government priorities and ensures that money is invested in the most valuable and viable projects.

An example of an action that was common to all water bodies was the management of wastewaters and water purity. In 2001, the project "Cueillir des mollusques aux Îles de la Madeleine" allowed the identification of contamination sources of several closed shellfish beds. This led to a better global comprehension of this complex problem that involves multiple government bodies monitoring the water quality, shellfish toxins, meat, etc. A group of all the stakeholders was formed, permitting the various departments involved to talk, exchange information and identify solutions. This process exposed the complex issue of exotic species introduction and started a conversation about what can be done to diminish this important risk to farming.

The benefits of this whole process are:

- Networking various interests with a simple approach;
- Avoiding fragmentation of activities;
- Maintaining activities while respecting users;
- Involving users;
- Respecting the interests of each sector;
- Minimal funding is needed for the management of the committee, lead by a local group.

Unfortunately in 2008, Fisheries and Oceans Canada stopped the funding for the integrated management that was taking place within different organizations of the Esturary and Gulf of St. Lawrence where this department had jurisdiction on its water under the Canadian *Oceans Act*. The provincial government under the "politique nationale de l'eau" enacted in 2002 followed by a provincial law in 2009 would give prominence to integrated management of water resources which lead to a framework on water management in 2012 (MDDEFP 2012). This repositioning permitted management to go beyond the salted water areas of the St Lawrence River to develop a new mode of global governance. However, the regions that relied on support from Fisheries and Oceans Canada had difficult years ahead because this important and effective mechanism for integrated management that animated integrated committees now depended entirely on volunteer efforts.

This situation resulted in the loss of some of the committees. There were a few other integrated management committes in the east coast... only those of the Magdalen Islands survived and their survival is said to be *"due to the direction and support the ZIP Committee of the Magdalen Islands gives to its actions"* (Steeve 2011) The ZIP Committee was doing its best to ensure that the integrated management committees only had to do minimal work to keep the lifeblood of the community flowing and to better prepare for what lay ahead.

Forty watershed organizations (OBV) have now developed along main tributaries of the St. Lawrence, and the last St. Lawrence agreement, in 2011 (http://planstlaurent.qc.ca/en/integrated_management/regional_round_tables.html), led to the development of regional roundtables (TCR, Table de Concertation Régionale) that would produce regional integrated management plans for twelve zones, representing the main sections of all the St. Lawrence. A framework was produced to ensure that practices would be harmonized and complementary (Plan d'action Saint-Laurent 2011-2026, présentation aux comités ZIP et à Stratégies Saint-Laurent, déc. 2011 Québec-Canada). (Figure 3). It has many advantages over the previous approaches to integrated management, the main one being that it takes place all along the St. Lawrence River and permits a global approach. A national forum takes place annually to ensure linkages and a common vision in all the processes.

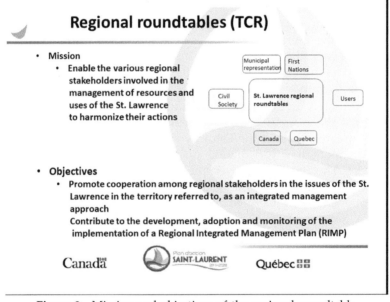

Figure 3   Mission and objectives of the regional roundtable.

In 2014, the Quebec Ministry of Environment (MDDELCC) and the community managers officially recognized the ZIP Committee of the Magdalen Islands as the primary entity responsible for the deployment of integrated management through the formation of a regional roundtable (TCR) for our Zone. A board of fifteen members comes from local organizations that have their "feet in the water". Together, they produced the diagnostic portrait of the zone (2017), for a better deployment of integrated management of fresh and salted waters which are very closely interrelated. An action plan will follow shortly. The website details the approach: http://zipdesiles.org/tcr/.

Meanwhile, much useful, but not always neutral information was shared in all the studies and processes needed for the development of a possible oil industry in the St. Lawrence. A local risk management center for possible maritime incidents was created in the context of the maritime strategy of the provincial government. A report on the state of the oceans for the integrated zone of the Gulf of St. Lawrence was also produced (Department of Fisheries and Oceans 2012) and helped to put into context the global phenomenon of, for example the effect of acidification, or reduced ice coverage during winter. Also, a project for a marine protected area around the islands creates opportunity for new studies and information to building objective and detailed portraits of all aspects of the zone. The Magdalen Islands are also studying the feasibility of gaining the status of a "biosphere reserve" from UNESCO to further secure sustainable development.

The Magdalen Islands archipelago depends on its coastal areas. Education provides hope for the Magdalen Islands and training individuals in the application of environmental risk management of coastal areas is a key component to ensuring the island sustainability into the future. The Magdalen Islands campus of the Cégep de la Gaspésie et des Îles recently developed a new program dedicated to better knowledge and more respectful interventions on the fragile coastal areas and its inland water bodies. To do this, it is important to share within the community, but also to have the support of the managers and the scientific community to better understand the changing environment and the new problems that have to be solved together. Then it might be possible to enjoy the resources in a sustainable way and live in harmony with one another and with the natural environment on the small territory for the longest time feasible.

REFERENCES

Agreement Canada-Quebec on the St. Lawrence. 1988. http://planstlaurent. qc.ca/en/integrated_management/regional_round_tables.html Accessed August 31 2014.

Bergeron, J. 2007. Strategy of implementation of the integrated management plans for the Magdalen Isalnds' five interior water bodies. Report prepared for Fisheries and Oceans Canada, Quebec Region. 78 p. http://www.zipdesiles. org/documents/ Accessed April 2014.

Department of Sustainable Development, the Environment, Wildlife and Parks. 2018. Integrated water resources management: terms of reference http:// legisquebec.gouv.qc.ca/en/pdf/cs/M-30.001.pdf Accessed November 2018.

Department of Fisheries and Oceans. 2012. State-of-the-Ocean Report for the Gulf of St. Lawrence Integrated Management (GOSLIM) Area http://publications. gc.ca/site/eng/428885/publication.html

Friesinger, S. and Bernatchez, P. 2010. Perceptions of gulf of St. Lawrence coastal communities confronting environmental change: hazards and adaptation, quebec, Canada. Ocean and Coastal Management. 53: 669-678.

MELCC. 2018. Integrated water resources management: terms of reference. http://www.environnement.gouv.qc.ca/eau/st-laurent/gestion-integree/ tables-cr.htm Accessed April 2014.

Municipalite des Iles-de-la Madeleine. 2018. Horizon 2025. Batir ensemble l'avenir. Un projet de territoire pour les Iles-de-la-Madeleine. https://www.muniles. ca/wp-content/uploads/Document_FINAL_PROJET_DE_TERRITOIRE.pdf Accessed November 25, 2018.

NASA. 2018. Global climate change http://climate.NASA.gov/key_indicators/ #seaLevel Accessed November 2018.

Plante, S. 2011. Les défis de la gestion intégrée des territoires côtiers et riverains du Saint-Laurent. pp. 110-117. *In*: O. Chouinard et al. (eds) Zones côtières et changement climatique : le défi de la gestion intégrée, Québec : Presses de l'Université du Québec.

Rémillard, A.M., St-Onge, G., Bernatchez, P., Hétu, B., Buylaert, J.-P., Murray, A.S. and Vigneault, B. 2016. Chronology and stratigraphy of the Magdalen Islands archipelago from the last glaciation to the early holocene: new insights into the glacial and sea-level history of eastern Canada. Boreas. 10.1111/bor.12179. ISSN 0300-9483.

St. Lawrence Action Plan 2011-2016. http://planstlaurent.qc.ca/en/home.htm Accessed August 1, 2014.

Chapter 4

Baynes Sound; Its unique nature and the need to recognize the region as a Marine Protected Area

Leah Bendell

Professor, Biological Sciences, Simon Fraser University, Burnaby, BC

BAYNES SOUND AS AN ECOLOGICALLY AND BIOLOGICALLY SENSITIVE AREA

Close to twenty years ago, the federal government adopted Canada's Oceans Act. Its purpose was to provide the legislative framework for an integrated ecosystem approach for the management of Canadian oceans inclusive of coastal zones (Department of Fisheries and Oceans [DFO] 2013). For British Columbia (BC), this region encompasses more than 27,000 kilometers of coastline and 105 river systems for which the federal DFO is responsible for management and protection. The Oceans Act states that "conservation, based on an ecosystem approach is of fundamental importance to maintaining biological diversity and productivity in the marine environment". Areas of Ecological and Biological Significance (EBSAs) had not yet been identified for coastal BC prior to 2012. In February of 2012, DFO provided an evaluation of proposed EBSAs for the marine waters of BC with the outcomes published as a Canadian Science Advisory report (DFO 2013).

The identification of EBSAs in the Canadian Pacific Region was to serve as a key component of the knowledge base for regional development activities and marine use planning, the development of Canada's network of Marine Protected Areas (MPAs) under the Oceans Act, and facilitating the implementation of DFO's sustainable fisheries framework. In addition, information generated from the identification of EBSAs was to be of value to other federal departments and the province of BC who are responsible for the management of marine activities within coastal BC (e.g. resource extraction, marine shipping, ocean dumping, spill response, cable laying, land use planning, etc.) (DFO 2013).

Core criteria for an EBSA include unique, rare, and distinct features, species aggregation, including areas where most individuals of a species are aggregated for some part the year, fitness consequences defined as areas that are used by species for life history activity(ies) and that make a significant contribution to the fitness of individuals of those species. An EBSA would meet one or more of these three core criteria. Baynes Sound, meets all of the criteria.

Baynes Sound is located within the biologically significant region of the Strait of Georgia (Chapter 1, Figure 1). Its physical features which include being a thermally stratified inland sea with soft substrate and biological fronts meet the unique physical criteria for an EBSA. Its uniqueness also includes its key location for marine birds and ranks only second to the Fraser River Delta as a region that supports globally significant numbers of migratory birds. (Chapter 6). Species that aggregate within this region include the butter clam (DFO 2013). Fitness consequences include the region being a staging area for marine birds (e.g., brant, harlequin ducks), a foraging area and haul out for Stellar sea lions and a key spawning and rearing area for forage fish.

Forage fish are small fish that are prey for thousands of species of birds, marine mammals and larger fish. Pacific herring, anchovy, Pacific sand lance, surf smelt, sardine, capelin and eulachon provide examples and occur in school sizes defined in metric tonnes. As a result of this huge biomass, forage fish are the most important link in the coastal food web converting zooplankton to protein for higher trophic levels, thus representing a "cornerstone" of the nearshore food web. Forage fish are key dietary items for BC salmon and comprise over 70% and 50% of the prey captured by Chinook and Coho salmon, respectively. Forage fish are also a major commercial fishery playing a significant economic role in coastal communities. Baynes Sound provides key intertidal habitat for all species of forage fish including sand lance and herring (deGraffe 2014).

Herring are probably the most abundant forage fish in BC (Hay 2014). They spawn in early spring, intertidally and in shallow subtidal waters and live eight to twelve years, reaching maturity at three years of age. Records of herring spawning locations in Baynes Sound and Lambert Channel starting in the 1930s, show about 38 percent of all the herring spawning that has ever occurred on the BC coast has occurred in two general areas: 35 percent in the Lambert Channel area and three percent in the areas immediately north of Cape Lazo. (Figure 1). Although herring spawn occurs within Lambert Channel, larvae are advected into Baynes Sound demonstrating the important ecological role that the Sound plays in providing forage fish habitat. Climate change has occurred within the Strait of Georgia and may be one of several reasons why spawning is increasingly occurring northward while decreasing southward. There may also be other reasons, however. Spawning herring will move if spawning habitats in the intertidal and subtidal locations have in some way been altered and become unsuitable e.g., as the result of anthropogenic factors such as oil spills, marinas and mariculture (Hay 2014).

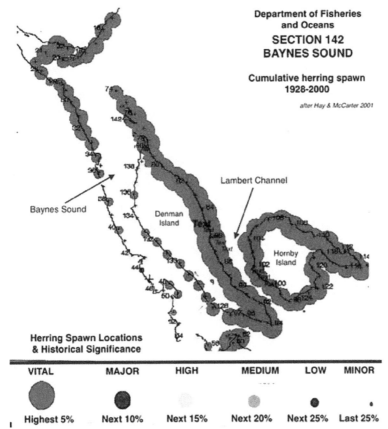

Figure 1 Herring spawn locations within Baynes Sound, Lambert Channel.
(For color image of this figure, see Color Plate Section at the end of the book)

These combined features make Baynes Sound one of the most unique and biologically sensitive regions along coastal BC. Despite its ecological importance, Baynes Sound is also a region under an increasing number of competing economic stressors including increased urbanization, seaweed harvesting (Chapter 7), an expanding shellfish aquaculture industry (Chapter 8), and pollution (Chapter 9). Of these, the unregulated shellfish industry is perhaps the greatest threat to the Sound.

THE SHELLFISH INDUSTRY WITHIN BAYNES SOUND

Shellfish aquaculture has a long and rich history on the west coast of British Columbia. At the turn of the 20th century, the shellfishery was a major attractant for immigrant families of all origins (BC Packers 2013). The species of shellfish harvested were native species, the Olympic oyster, and littleneck, butter and razor clams. With the inevitable overexploitation of the native species came the import of non-indigenous species, first the Atlantic and then the Pacific oyster, the latter which is now the foundation of the oyster industry worldwide.

In British Columbia, the unregulated introduction of the Pacific oyster occurred ca. 1912 (reviewed in Gillespie et al. 2012). By 1942, widespread reproductive success had occurred leading to the establishment of populations throughout the Strait of Georgia (Gillespie et al. 2012). The year 1948 saw the establishment of the BC Oysters Grower's Association which became the BC Shellfish Growers Association (BCSGA) in 1990 to better reflect the increasing diversity of the industry.

In 1997, the federal department of Western Economic Diversification commissioned a study by Coopers and Lybrand (Coopers and Lybrand 1998), to review the economic potential of the BC marine aquaculture industry. The study found that shellfish farming had the potential to become a $100 million dollar industry that would create more than 1,000 person years of employment in BC coastal communities over the next ten years. This potential was based on the presence of significant long-term market opportunities, inventories of capable aquatic lands, technological improvements and development of new shellfish species. Access to new farm sites was identified as the greatest factor limiting the industry. In November 1998, the provincial government announced a Shellfish Development Initiative to double the Crown land available for shellfish aquaculture at a rate of 10% per year over ten years. In 1998, the amount of area committed to the shellfishery was 2300 ha. with 427 tenures. In 2010-2012 the amount of area was 3500 ha. with 467 leases (Mamoser 2011; Vann Struth 2014). In 2016, DFO reported approximately 450 shellfish tenures with 3800 ha. under lease (Shellfish Integrated Management and Aquaculture Plan (SIMAP) 2017). In 2018 there were 417 tenures of which 137 were located in Baynes Sound (Figure 2) (DFO 2018). Of these 137, 33% are owned by four operators, with Taylor Shellfish and Mac's Oysters owning 10% and 12% of the total number of leases (DFO 2018). Shellfish aquaculture employs 20% of the BC aquaculture sector or 320 jobs per year, with approximately 100 people within the Baynes Sound area (SIMAP 2017). In 2015, British Columbia produced 10,550 tonnes of shellfish of which 63% were oysters. Oysters continue to account for the majority of wholesale value averaging 51% over the past ten years from 2006 to 2016 with the majority of production occurring within Baynes Sound. Currently 90% of the intertidal region of the Sound is under shellfish lease (Figure 2) with the majority (80%) of the major product, oysters, sold to US markets (Kitchen 2011).

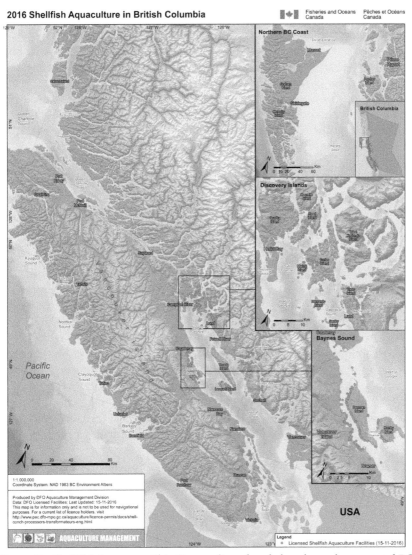

Figure 2   Shellfish tenures within Baynes Sound and elsewhere along coastal BC.
(For color image of this figure, see Color Plate Section at the end of the book)

BAYNES SOUND; THE NEED TO ESTABLISH THIS REGION AS A MARINE PROTECTED AREA

Although the expansion of the shellfish industry has fallen well short of expectations, the potential for the rapid expansion of the industry notably within Baynes Sound triggered the Fisheries Act subsection 35 (2) relating to the Harmful Alteration, Disruption and Destruction of fish habitat (HADD). The concern raised prompted a Phase 0 habitat review of the Sound, which was to provide a baseline, such that advice on alternative management options could be put in place, and to identify where information on existing shellfish practices was lacking (Jamieson et al. 2001). Included in the key recommendations from the 2001 Phase 0 review (of which none were acted on) were: 1) the establishment of a multi-agency research initiative **as soon as possible.** This would include Environment Canada avian researchers, DFO marine invertebrate and fish experts, the Geological Survey of Canada and the Canadian Hydrographic Service to identify both the nature of existing and potential impacts and determine how they could be minimized, and 2) establishing an effective network of protected areas in Baynes Sound that would exclude shellfish culture to be established. It was **also noted that the significance of Baynes Sound in the Georgia Basin ecosystem had not been recognized by resource managers to date.** None of these recommendations have since been implemented.

Important was recommendation 2 which in 2001 stated that an effective network of protected areas within Baynes Sound should be established. This was prior to the 2012 DFO report that identified Baynes Sound as one of the most important ecosystems along the Pacific Northwest coast. As noted by Jamieson et al. in 2001, there has been no attempt by the federal government to acknowledge the ecological importance of Baynes Sound, despite the recommendations and warnings from its own department, DFO. Rather, the unregulated expansion of the shellfish industry continues with an increasing number of leases being approved for the farming of geoducks (DFO 2017) to meet the strong demand by the Asian markets for a luxury protein. And with it, are the increasing cumulative impacts that the industry practices have on the Sound (Bendell 2015, Chapter 8).

Current practices employed by the BC shellfish industry include: the extensive use of anti-predator netting within the intertidal to prevent predation of clams by avian predators, extensive use of the foreshore to "grow out" oysters prior to market sale, extensive use of plastics in all equipment such as trays, rope, netting, buoys, fencing, oyster socks and PVC piping for geoduck seed (Chapter 8), driving on the foreshore leading to the loss of forage fish habitat, loss of eelgrass and all the ecological services associated with eelgrass habitat, and extensive use of rebar resulting in beach hazards. Equipment used by the industry is not secured and quickly becomes derelict fishing gear which washes up onto the shore. Since 2004, the local residents of Baynes Sound have held an annual beach cleanup; from 2004 to 2010 approximately 2-3 tonnes of shellfish debris was collected. In 2018, over 6 tonnes was collected despite the continued efforts of the local community to alert DFO to the amounts of plastic debris that was originating from the industry (Figure 3).

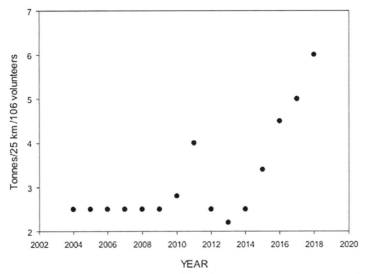

Figure 3   Tonnage of debris collected over a 14 year time period. Data courtesy of ADIMS (Association for Denman Island Marine Stewards)

Perhaps what is the most perplexing is the absence of response and the denial by both the industry and DFO of the mismanagement and complete lack of respect for Baynes Sound, a region that has been identified by DFO as an EBSA and recommended by DFO over 17 years ago to be recognized as an MPA. Indeed, the view of the industry is quite the opposite as indicated by a recent interview given by the current executive director of the BCSGA who continues to call for expansion despite the negative impacts that industry is having. The executive director points to the regulatory process for assigning modifying or expanding tenures as the cause for absence of growth in the shellfish aquaculture sector (Aquaculture North America 2018). She also notes that the vast majority of their farmers are quite responsible "whether they're on the farm or travelling back and forth, debris is an issue. They stop and pick it up ... we encourage our members to be good neighbors". The amount of debris that has been collected from the beaches of Baynes Sound has tripled since 2010 (Figure 3) and does not support this description of a responsible industry.

SHELLFISH; CANARIES OF THE SEA

However, despite the optimistic outlook of the shellfish industry, it is not the access to increased tenures that is restricting the growth of the industry. The shellfish industry is facing an uncertain future given all that we have done to our oceans over the last 70 years. Shellfish are intimately associated with the water column, requiring clean cool waters rich in algae to succeed. Anthropogenic activities that compromise this environment will negatively impact shellfish, both wild and farmed. The World Economic Forum (2017) recently compiled 25 tipping points pushing our oceans past the point of no return. Included in the 25 were ocean acidification, rising ocean temperatures resulting in the increase of harmful algae blooms and shellfish borne diseases, plastics pollution and more intense storms. Shellfish on the Pacific west coast are facing all of these challenges now.

A recent review by Haigh et al. (2015) highlighted the vulnerability of shellfish aquaculture to Ocean Acidification (OA) noting that " shellfish aquaculture is highly susceptible to OA due to the direct impact of OA on shell formation and the dependence of the industry on hatchery production". Moore-Maley et al. (2017) also note that the growth of the BC shellfish industry has fallen behind 1997 projections and identify changes in carbonate chemistry due to ocean acidification which reduces growth and survival in juvenile shellfish as a possible explanation. They reported that the Salish Sea is acidic relative to other regions of the northeast Pacific Ocean and seawater below 20m is potentially corrosive to shellfish. This corrosive seawater flows into Baynes Sound as a result of south winds from the Salish Sea.

Within Baynes Sound, pH hovers around 7.6 and can go as low as 7.2 (http://nvs.nanoos.org/Explorer?action=oiw:fixed_platform:FBO_Fannybay1:observations:H2_CO2). Indeed, the impacts of ocean acidification are already being experienced in BC. In 2014, the loss of 10 million scallops and a 90% mortality of oysters in coastal BC waters were attributed to OA (https://www.theglobeandmail.com/news/british-columbia/mystery-surrounds-massive-die-off-of-oysters-and-scallops-off-bc-coast/article17156108/).

Rising ocean temperatures which trigger Harmful Algae Blooms (HABs) and shellfish borne diseases also plague the industry. Since 1984, there has been an increased frequency, intensity and geographical distribution of HABs (Hallengraeff et al. 2004). A recent study by Lefebvre et al. (2016) concluded that current climate trends which cause rapid declines in sea ice and increasing water temperatures are likely to expand the northern geographic range and duration of conditions that will facilitate HABs. The year 2015 saw possibly the largest bloom ever detected, one that stretched from California north to BC. (https://www.cbc.ca/news/canada/british-columbia/toxic-algae-bloom-off-west-coast-might-be-largest-ever-1.3116914).

More recently, the BC shellfish industry experienced a 9.1 million dollar loss due to closures as a result of norovirus outbreak (Miller et al. 2018). The outbreak affected over 400 Canadians over the period of late 2017 to early 2018. Six of the twelve farms closed were located in Baynes Sound. While multiple sources of human sewage entering the marine environment was identified as the likely causative agent, Miller et al. (2018) also note that the outbreak likely occurred due to favorable environmental conditions that would facilitate the transport to and uptake and accumulation by oysters. Changing climate conditions was also linked to the ability of the norovirus to survive. Norovirus is seasonal with most outbreaks occurring during winter. Hence, Miller et al. 2018 suggest that heavy rainfall and colder than normal temperatures allowed norovirus to persist in marine waters. They also warn that another outbreak is highly likely. Finally, oysters from Baynes Sound have on average 2.4 ppm (range 1.17-3.57 ppm) of cadmium, a toxic element, present in their tissues (Lekhi et al. 2008). Because of this amount, they cannot be sold to either the European or Hong Kong markets as amounts present exceed export market health safety guidelines of 1 and 2 ppm.

Perhaps the greatest challenge and a newly identified threat to a struggling shellfish industry is that posed by microplastics. Microplastics defined as particles < 5 mm diameter can include fibers (or filaments), fragments and spheres (Kazmiruk et al. 2018). The presence of both macro and microplastics within our ocean ecosystems has been identified as one of the greatest threats that ocean life has faced (Hurley et al. 2018). Kazmiruk et al. (2018) recently discovered that Baynes Sound was highly contaminated with microplastics, notably microbeads. Given the amounts of plastic debris collected each year by the local Denman Island community volunteers, perhaps this finding was not unexpected as plastics such as polypropylene ropes and polyethylene terephthalate are extensively used by the industry. What was surprising were the numbers and mass of microbeads found in the surface sediments in the north end of the Sound. Concentrations were equivalent to those of natural sediments components such as silt and organic matter.

Recently, Murphy (2018) compared numbers of microplastics in three types of shellfish, mussels, clams and oysters both farmed (purchased) and wild collected from around Vancouver Island including regions close to Baynes Sound. Both farmed mussels and oysters contained significantly greater numbers of microplastics as compared to their wild counterparts with 99% of particles identified as microbeads. Recent studies of Sussarellu et al. (2016) have shown that oyster reproduction is adversely affected by exposure to polystyrene microbeads by interfering with energy uptake and allocation, reproduction, and overall performance. Whether the presence of microbeads within Baynes Sound is having such an affect on farmed and wild oysters is at this time unknown.

CONCLUSIONS

In 2001, close to 20 years ago a Phase 0 review conducted by the Federal Department of Fisheries and Oceans on the potential impacts of an expanding shellfish industry within Baynes Sound recommended that regions within the sound be set aside as MPAs. The ecological and biological importance of Baynes Sound was again indicated by DFO in 2012 when specific regions within the Salish Sea were identified as EBSAs or Ecologically and Biologically Sensitive Areas. Despite this advice, the shellfish industry in collaboration with DFO continues to negatively impact the Sound with an ever-increasing tonnage of plastic debris accumulating on the foreshore with even further plans for expansion.

The industry faces an uncertain future. Climate change is resulting in warming oceans with increased occurrences of toxic algae blooms. The oceans are acidifying leading to massive shellfish die offs. Changing oceanic conditions due to climate change are facilitating the increased indices of viruses, such as the norovirus which lead to a 9.1-million-dollar loss to the industry in 2017. BC oysters contain 1- 4 ppm cadmium, a toxic element, concentrations that preclude the selling of the product to European and Asian export markets (Bendell, 2013). And now we are finding that oysters also contain microplastics and we do not yet know what the human health implications are.

Baynes Sound is likely the most important inland sea ecosystem along coastal BC. It needs to be managed as a marine protected area, not as a region for the unregulated expansion of an industry that is producing one product which is a luxury protein for an Asian market and another product which contains cadmium and microplastics for an US market who are mostly males consuming oysters at local oyster bars (Bendell 2013).

What shellfish farming will look like in 10 years from now is uncertain. By contrast, we can act now to restore and protect Baynes Sound as a Marine Protected Area with a known outcome. In closing, in and above protecting all of the unique features of the Sound that makes this region an EBSA, the restoration and protection of our coastal ecosystems is paramount for our own survival.

REFERENCES

Aquaculture North America. 2018. https://www.aquaculturenorthamerica.com/shellfish/bc-shellfish-farmers-set-for-another-year-of-growth-1914. Accessed November 1, 2018.

BC Packers. 2013. http://www.intheirwords.ca/english/people.html. Accessed June 2013.

Bendell, L.I. 2013. The cadmium conundrum in British Columbian oysters: Economics, ecotoxicology and ethics. Invited book chapter. *In*: Oysters: Biology, Consumption and Ecological Importance. Nova Publishers.

Bendell, L.I. 2015. Favored use of anti-predator netting (APN) applied for the farming of clams leads to little benefits to industry while increasing nearshore impacts and plastics pollution. View Point. Marine Pollution Bulletin. 91: 22-28.

Coopers and Lybrand. 1998. Economic potential of the British Columbia aquaculture industry: shellfish, Report Prepared for Western Economic Diversification, 1997.

deGraffe, R. 2014. Stewarding the sound; the challenge of managing sensitive ecosystems. Conveners Report. https://www.sfu.ca/content/dam/sfu/coastal/Research%20%26%20Series/Series/Management%20of%20Sensitive%20Ecosystems/StewardingTheSoundFinal.pdf. Accessed October 3, 2018.

DFO. 2013. Evaluation of proposed ecologically and biologically significant areas in marine waters of British Columbia. DFO Canadian Science Advisory Secretariat Science Advisory Report. 2012/075.

DFO. 2017. Integrated Geoduck Management Framework 2017 Pacific Region. http://waves-vagues.dfo-mpo.gc.ca/Library/40596862.pdf

DFO. 2018. Current valid British Columbia Shellfish Licences. Aquaculture Management. https://open.canada.ca/data/en/dataset/522d1b67-30d8-4a34-9b62-5da99b1035e6 Accessed October 28 2018.

Gillespie, G.E., Bower, S.A., Marcus, K.L. and Kieser, D. 2012. Biological synopses for three exotic mollusks, V. philippinarum (*Venerupis philippinarum*), Pacific Oyster (*Crassostrea gigas*) and Japanese Scallop (*Mizuhopecten yessoensis*) licensed for aquaculture in British Columbia. DFO Canadian Science Advisory Secretariat Research Document. 2012/2013. v 97p.

Haigh, R., Ianson, D., Holt, C.A., Neate, H.E. and Edwards, A.M. 2015. Effects of ocean acidification on temperate coastal marine ecosystems and fisheries in the northeast pacific. PLoS ONE 10(2): e0117533. doi:10.1371/journal.pone.0117533

Hay, D. 2014. Stewarding the Sound; the challenge of managing sensitive ecosystems. Conveners Report. https://www.sfu.ca/content/dam/sfu/coastal/Research%20%26%20Series/Series/Management%20of%20Sensitive%20Ecosystems/StewardingTheSoundFinal.pdf Accessed October 3, 2018.

Hallegraeff, G.M., Anderson, D.M. and Cembella, A.D. 2004. Manual on Harmful Marine Microalgae. United Nations Educational, Scientific and Cultural Organization.

Hurley, R., Woodward, J. and Rothwell, J. 2018. Microplastic contamination of river beds significantly reduced by catchment-wide flooding. Nature Geoscience 11: 251-257.

Jamieson, G.S., Chew, L., Gillespie, G., Robinson, A., Bendell-Young, L., Heath, W., Bravender, B., Nishimura, D. and Doucette, P. 2001. Phase 0 review of the environmental impacts of inter-tidal shellfish aquaculture in Baynes Sound. Fisheries and Oceans Canada, Canadian Science Advisory Secretariat, Ottawa, Ontario. 103 p.

Kazmiruk, T.N., Kazmiruk, V.D. and Bendell, L.I. 2018. Abundance and distribution of microplastics within surface sediments of a key shellfish growing region of Canada. PLoS ONE 13(5): e0196005. https://doi.org/10.1371/journal.pone.0196005

Kitchen, P. 2011 An economic analysis of shellfish production associated with the adoption of integrated multi-trophic aquaculture in British Columbia. MRM, School of Resource and Environmental Management, Simon Fraser University, Burnaby BC.

Lekhi, P., Cassis, D., Pearce, C.M., Ebell, N., Maldonado, M.T. and Orians, K.J. 2008. Role of dissolved and particulate cadmium in the accumulation of cadmium in cultured oysters (*Crassostrea gigas*). Sci Total Environ 393: 309-325.

Lefebvre, K.A., Quakenbush, L., Frame, E.R., Burek-Huntington, K., Sheffield, G., Stimmelmayr, R., Bryan, A., Kendrick, P., Ziel, H. and Goldstein, T. 2016. Prevalence of algal toxins in Alaskan marine mammals foraging in a changing arctic and subarctic environment. Harmful Algae, 55: 13-24.

Mamoser, M.P. 2011. Towards Ecosystem-Based Management of Shellfish Aquaculture in British Columbia. Canada, An industry perspective. MSc. University of Victoria. 178 pp.

Miller, A., Cumming, E., McIntyre, L. and the Environmental Transmission of Norovirus into Oysters working group members. 2018. Summary working group report of the environmental transmission of norovirus into oysters following the 2016-17 national outbreak of norovirus linked to the consumption of BC oysters. Environmental Health Services, BC Centre for Disease Control. June 2018.

Moore-Maley, B., Ianson, D. and Allen, S. 2017. Wind-Driven upwelling and seawater chemistry in British Columbia's shellfish aquaculture capital. *In*: Oceans, What's Current. http://bulletin.cmos.ca/upwelling-seawater-chemistry-bc-shellfish-aquaculture/ Accessed November 5, 2018.

Murphy, C. 2018. A comparison of Microplastics in Farmed and Wild Shellfish near Vancouver Island and Potential Implications for Contaminant Transfer to Humans. MSc thesis. Royal Roads University, Victoria BC. 75 pp.

Sussarellu, R., Suquet, M., Thomas, Y., Lambert, C., Fabioux, C., Pernet, M.E.J. Goïc, N.L., Quillien, V., Mingant, C., Epelboin, Y., Corporeau, C., Guyomarch, J., Robbens, J., Paul-Pont, I., Soudant, P. and Huvet, A. 2016. Oyster reproduction is affected by exposure to polystyrene microplastics. Proceedings of the National Academy of Sciences 113: 2430-2435.

Vann Struth Consulting Group. 2014. Aquaculture Statistics for Comox Valley and BC. 3 pp. https://discovercomoxvalley.com/wp-content/uploads/2016/12/Aquaculture StatisticsforComoxValley-1.pdf Accessed November 1, 2018.

World Economic Forum. 2017. https://www.weforum.org/agenda/2017/06/25-ocean-tipping-points/ Accessed October 3, 2017.

Chapter 5

Overview of Baynes Sound Salmonids and Possible Limiting Factors Important for Local Ecosystem Management

Colin Levings

Scientist Emeritus (Retired),
Department of Fisheries and Oceans,
Centre for Aquaculture and Environmental Research,
4160 Marine Drive, West Vancouver, BC, Canada V7V 1N6

INTRODUCTION

The objective of this paper is to provide a brief overview of our current knowledge of the estuarine habitat ecology of salmonids in Baynes Sound, including comments on possible limiting factors. The Sound has been long recognized as an important salmonid production area (Morris et al. 1979). Twenty-three rivers and creeks on the west side of the Sound, as well as the Courtenay River, draining into the Sound at its north end, provide extensive spawning and rearing habitat. The estuarine and marine shorelines are used as rearing habitat for all of the anadromous salmonid species found in British Columbia (Hamilton et al. 2008), except for Dolly Varden Char and bull trout. The offshore of the Sound is likely a major migratory corridor for juvenile salmonids moving seaward (e.g. chum and pink salmon, Fraser et al. 1979) from the Fraser River and other rivers on the southern Strait of Georgia. Thus a range of Sound habitats are essential to maintain salmonid diversity and production.

LIFE HISTORY OF BAYNES SOUND SALMONIDS

Anadromous salmonids are hatched from eggs spawned in freshwater, move to the sea where they attain sexual maturity, and then move back to freshwater to reproduce. Anadromous salmonids display two forms of reproduction or life history plans (semelparity and iteroparity) which determine their degree of estuarine use. Semelparous species migrate from the sea to spawn in fresh water and then die. Documented examples from Baynes Sound are chum salmon (*Oncorhynchus keta*), chinook salmon (*Oncorhynchus tshawytscha*), coho salmon (*Oncorhynchus kisutch*), sockeye salmon (*Oncorhynchus nerka*) and pink salmon (*Oncorhynchus gorbuscha*). Most iteroparous species spawn and overwinter in rivers but migrate into estuaries or adjacent coastal waters for several summers to feed but eventually die in the river after a final spawning. Over their lifetime, iteroparous species obviously spend more of their life in estuaries relative to semelparous species. Examples from Baynes Sound are cutthroat trout (*Oncorhynchus clarki*) and steelhead *Oncorhynchus mykiss*).

DISTRIBUTION AND SEASONAL ABUNDANCE OF JUVENILE SALMONIDS IN THE SOUND

Data on the distribution and abundance of juvenile salmonids in the Sound are sparse, somewhat dated, and focus on the northwest sector of the Sound, with an emphasis on the sampling near the Courtenay River estuary. Comprehensive data are available for the juvenile salmonid ecology of the Courtenay River estuary proper (Lake Trail 2011).

The most recent comprehensive survey data for Baynes Sound, or at least the northwest portion of the Sound, are available for March to August 2001 and are given by Hamilton et al. (2008). Surveys by Hamilton et al. (2008) used beach seines and Gee trap to sample shallow water habitats and purse seines to cover deeper water. In this paper I focus on the beach seine data because of local interest on the nearshore.

Young salmon were found in estuarine beach habitats in the Sound from March until August and possibly later. Chinook salmon fry and smolts catches peak in June, while maximum chum salmon catches were in March (Figures 1-3). Most of the Chinook salmon may originate from the Courtenay River system, while chum salmon are likely from local rivers and creeks as well as rivers outside the Sound. Most of the coho salmon smolts were caught between May and June. Pink salmon were found only in March. Cutthroat trout were sampled mainly in June and July.

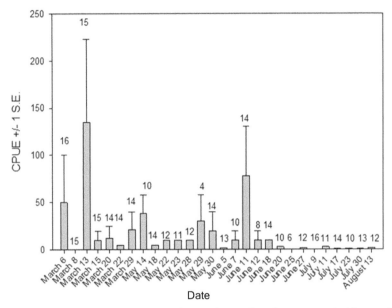

Figure 1 Mean number of salmonids caught per beach seine set (with standard error) in Baynes Sound from March to August 2001 (number of sets shown above bars or for specific dates) (from Hamilton et al. 2008).

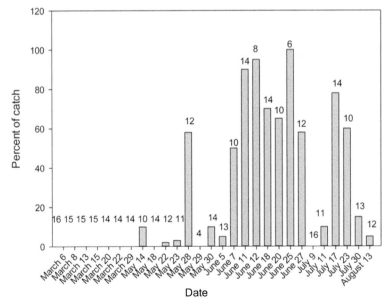

Figure 2 Catch of chinook salmon as a percent of total catch by date for all beach seine sites (number of sets shown above bars or for specific dates) (from Hamilton et al. 2008).

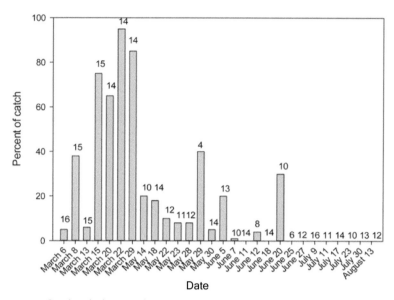

Figure 3 Catch of chum salmon as a percent of total catch by date for all beach seine sites (number of sets shown above bars or for specific dates) (from Hamilton et al. 2008).

HABITATS AND WATER PROPERTIES

Baynes Sound shoreline habitats include brackish marsh, mudflats and shrubs near the creek and river mouths, eelgrass meadows, kelp beds, algal mats, and salt marshes in higher salinity areas. Unvegetated substrates such as sand, gravel, and rocky areas are also widespread (Levings et al. 1999; Hamilton et al. 2008). Analyses of the distribution and abundance of juvenile salmonids in relation to these specific shallow water habitat types are not available for the Sound. Hamilton et al. (2008) interpreted the distribution patterns they found on the basis of salinity zones. Six of the nine zones ranged between about 20-25 psu –these were areas further away from the mouth of the Courtenay River. Salinity at other three zones, located closer to the Courtenay River, ranged between zero and ten psu. Relationships between salinity zones and salmonid catches were variable and difficult to interpret because three different gear types were used. Further analyses of the catch data in Hamilton et al. (2008) could assist in identifying shoreline habitats for salmonids although there may be problems because of differing sampling efficiency on the various habitat units.

The Sound proper might be considered a multiple estuary because the numerous creek and river mouths have their own local salinity. The numerous rivers and small streams have widely varying discharge rate. Maximum flows are frequently in both spring and fall, driven by rainfall, but with low summer flows (Figure 4). Considering offshore water masses, the north end of the Sound is characterized by lower salinity because of the influence of the Courtenay River (Hamilton et al. 2008). Salinities in the Sound vary seasonally, with depth, with tidal stage, with the distance from the local freshwater discharge, and with water masses intruding from the Strait of Georgia (Masson and Peña 2009). Effects can be localized. Surface salinity in Deep Bay (water depth 25 m, a few hundred metres offshore of the mouth of Cook Creek) ranged from about 23-29 psu and was highest in winter (Cassis et al. 2011). Salinities at creek mouths are not available but would be expected to range down to zero psu. Lekhi et al. (2008) found surface temperature in Deep Bay to range from 7 to 19 °C, likely representative of offshore in the Sound. Temperature in the immediate vicinity of creek mouths would range lower (e.g., down to 2 °C in winter at Lymn Creek, a small stream entering Deep Bay (Mason 1974a)).

THE SALMONID RIVER-ESTUARY-OCEAN CONTINUUM IN THE SOUND

We know from studies elsewhere in the Strait of Georgia that juvenile salmonids use a series of linked habitats as they move seaward – this can be described as the river-estuary-ocean continuum. The continuum can be tracked by salinity changes and biota shifts related to salinity differences. Data from a study on the rearing and feeding ecology of chum salmon fry in Lymn Creek and its estuary (Mason 1974b) is illustrative. Although dated, the information is representative of the conditions almost 50 years ago when Baynes Sound watersheds and estuaries were likely less disrupted relative to today.

In the autumn, chum salmon spawn in the lower reaches, and sometimes in the estuary of Lymn Creek (Mason 1974a). Their eggs incubate in the stream bed until spring when the fry downstream move downstream to the estuary. Chum salmon fry caught in Lymn Creek estuary ranged in length from about 30 to 55 mm in May (Mason 1974b), and showed most of their growth in the estuary. The diet of chum salmon fry in the estuary showed a transition from freshwater prey needing adequate water flow conditions (e.g. stream insects) to estuarine invertebrates adapted to live on mudflats or marsh habitat (e.g. gammarid amphipods) to pelagic copepods required good water quality (e.g. calanoid copepods). Stomach contents likely reflected availability of the various prey, depending particularly on tidal stage. For example, pelagic calanoid copepods became available to the chum salmon fry at high tide when the water from offshore flooded into the creek mouth. As they progress through their life history the chum salmon fry are able to sequentially exploit food sources from the stream, estuary, and offshore habitat. Estuary food sources provide energy supplies for chum fry growth in the nearshore region. Subsequently the chum salmon fry move into deeper water in main Sound, the adjacent Strait of Georgia (Fraser et al. 1979), and eventually the Pacific Ocean, where they grow to adults.

BAYNES SOUND LIMITING FACTOR ANALYSIS, ECOLOGICAL INTEGRITY, AND RESILIENCE

As juvenile salmonids move through the river-estuary-continuum, they are challenged by multiple, and interacting, possible limiting factors (Figure 4, Table 1). Freshwater flow, water quality, provision of food

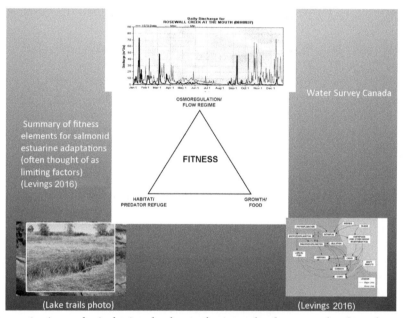

Figure 4 An ecological triangle shows the interplay between the three factors thought to relate to salmonid fitness in estuaries: freshwater flow and osmoregulatory affects; food supply and growth; and habitat as refuge (Levings 2016). The various species of salmonids in Baynes Sound can be aligned on the axes depending on their relative adaptiveness to each of three factors. Top corner graph: typical seasonally pattern of flow in Rosewall Creek (data for 1978 shown) ($m^3 \cdot s^{-1}$) (Water Survey of Canada 1978); lower left photo: brackish marsh habitat type (Courtenay River estuary; from Lake Trail 2011); lower right diagram: typical estuary food web (from Levings 2016).

Table 1 Possible limiting factors for juvenile salmonids in Baynes Sound and challenges presented by human alteration of watersheds, estuaries, and pelagic habitats

Possible limiting factor	Challenge	Example of potential problem in the sound
Osmoregulatory success	Provision of sufficient freshwater flows to maintain a freshwater-saltwater transition; water quality must also be maintained	Below optimum baseline flows in the Tsable River (BCCF 2014);
Food supply and growth	Maintenance of food production along the river-estuary-ocean continuum	Chum salmon fry feeding habitat in Lymn Creek estuary (Mason 1974a) and off Chrome Island, south end of Denman Island (Fraser et al. 1979) need to be maintained
Habitat refuges	Maintenance of shallow water and vegetated habitat for possible protection from predation	Sea run cutthroat trout entered the Lymn Creek estuary from November to March (Mason 1974b) indicating both the Sound and the Creek were winter habitat refuges (sensu Northcote 1997); shallow water and vegetated stream/estuarine habitat needs to be maintained

from estuarine habitat and habitat refuges are some of the important features that need to be maintained in the Sound and its estuaries. The maintenance and recovery of salmonid ecosystem properties such as ecological integrity and resilience could be a central management theme for the Sound. Ecological integrity is said to exist when an ecosystem is deemed characteristic for its natural region and resilience is said to be the amount of disturbance that an ecosystem could withstand without changing self-organized processes and structures. The resilience of the Sound's estuarine salmonid ecosystem is likely dependent on habitat, diversity and biological diversity at several levels (species, life history, genetic), but assessing these metrics is not easy. As well, resilience has to include the human socio-economic system with its inevitable ties to the salmonid ecosystem and its habitats.

Accounting for minor life history types of the various species of salmonids in the Sound is one possible tactic to assist in maintenance of resilience. The estuary-rearing coho salmon fry found in the Courtenay River watershed by Lake Trail (2011) and Tyron (2012) are an example.

Most of the coho salmon fry that survive to adults are thought to require freshwater residence for a year or two but detailed studies are revealing the importance of the estuary-rearing form (Tyron 2012). Other examples include the complex of estuary-rearing Chinook salmon life history types in the Courtenay River system (Guimond and Withler 2007) and the Big Qualicum River watershed (Lister and Walker 1966). Detailed ecological and genetic studies on juvenile salmonids utilizing the Sound's estuaries are required to increase our knowledge of resilience if life history types are to be used as a measure.

CONCLUSIONS AND RECOMMENDED STUDIES

The estuarine ecology of Bayne Sound salmonids is poorly known and ecosystem management of the area would benefit from further detailed studies, using an integrated watershed-estuary-ocean approach. Freshwater flow may be affected by extraction for human consumption, irrigation or other uses (Morris et al. 1979; Braybrook et al. 1995), as well as by climate change (Okey et al. 2014). Reduced flow could result in the development of temporary estuaries that are only present irregularly, with possible implications for salmonids as they move from freshwater to salt water. Nearshore habitat loss from urban development or past industrial activity (e.g. coal slag at Union Bay, Ryan 2008) has altered marine riparian vegetation important in food webs (Romanuk and Levings 2005). Possible short term effects of shellfish aquaculture on food webs (predator nets; Simenstad and Fresh 1995; Vandermeulen et al. 2006) and fish community issues (Bourdon 2015) are also topics that need resolving. As juvenile salmonids move into their pelagic habitats in the offshore and deeper areas of the Sound, temperature changes from climate modification may modify food web structure by incursion of invasive species (Levings et al. 2004) or water quality changes (warming from climate change; Okey et al. 2014). Thus a complex of factors could influence survival and further research is required to improve our understanding of anthropogenic effects.

As a first step in future salmonid ecosystem research, it would useful to compile a complete summary of the ecosystem status of Baynes Sound. A compilation of ecological data would be a necessary prelude to development of an ecosystem model incorporating food webs and anthropogenic changes. An update and expansion of the summary of environmental information presented by Morris et al. (1979) might be a template to follow.

REFERENCES

BCCF (British Columbia Conservation Foundation). 2014. Trent and Tsable Watersheds. http://www.bccf.com/steelhead/focus5.htm#tt Accessed June 10, 2014)

Bourdon, R. 2015. Interactions between fish communities and shellfish aquaculture in Baynes Sound, British Columbia. MSc Thesis, University of Victoria, Victoria, BC Canada. 98 p.

Braybrook, C., Bryden, G. and Dambergs, A., 1995. Nile creek to Trent River. Water allocation plan. Vancouver Island Region, Nanaimo, BC: Regional Water Management. 93 pp.

Cassis, D., Lekhi, P., Pearce, C.M., Ebell, N., Orians, K. and Maldonado, M.T. 2011. The role of phytoplankton in the modulation of dissolved and oyster cadmium concentrations in Deep Bay, British Columbia, Canada. Science of the Total Environment, 409(20): 4415-4424.

Fraser, F.J., Berry, S.J. and Allen, B. 1979. Big Qualicum River salmon development project. Volume IV, Chum fry marine study. Fisheries and Marine Service Technical Report 824.

Guimond, E. and Withler, R. 2007. Puntledge River summer chinook DNA analyses 2006. Comox Valley Project Watershed Society. P.O.Box3007, Courtenay, BC V9N 5N3.

Hamilton, S.L., Bravender, B.A., Beggs, C. and Munro, B. 2008. Distribution and abundance of juvenile salmonids and other fish species in the Courtenay River estuary and Baynes Sound, 2001. Canadian Technical Report of Fisheries and Aquatic Sciences No. 2806.

Lake Trail. 2011. Investigation of restoration and protection options for juvenile salmonids in the Courtenay River estuary. Comox Valley Project Watershed Society. P.O.Box 3007, Courtenay, BC V9N 5N3.

Lekhi, P., Cassis, D., Pearce, C.M., Ebell, N., Maldonado, M.T. and Orians, K.J. 2008. Role of dissolved and particulate cadmium in the accumulation of cadmium in cultured oysters (*Crassostrea gigas*). Science of the Total Environment 393: 309-325.

Levings, C.D., North, M.S., Piercey, G.E., Jamieson, G. and Smiley, B. 1999. Mapping nearshore and intertidal marine habitats with remote sensing and GPS: the importance of spatial and temporal scales, Oceans '99. MTS/IEEE. Riding the Crest into the 21st Century. Conference and Exhibition. Conference Proceedings (IEEE Cat. No.99CH37008), Seattle, WA, vol. 3: 1249-1255. doi: 10.1109/OCEANS.1999.800170.

Levings, C.D., Cordell, J.R., Ong, S. and Piercey, G.E. 2004. The origin and identity of invertebrate organisms being transported to Canada's Pacific coast by ballast water. Canadian Journal of Fisheries and Aquatic Sciences 61: 1-11.

Levings, C.D. 2016. Ecology of Salmonids in Estuaries Around the World: Adaptations, Habitats and Conservation. University of British Columbia Press. 388 p.

Lister, D.B. and Walker, C.E. 1966. The effect of flow control on freshwater survival of chum, coho, and Chinook salmon in the Big Qualicum River. Canadian Fish Culturist 37: 3-25.

Mason, J.C. 1974a. Movements of fish populations in Lymn Creek, Vancouver Island: a summary from weir operations during 1971 and 1972, including comments on species life histories Technical Report Canada. Fisheries and Marine Service. Research and Development Directorate 483.

Mason, J.C. 1974b. Behavioral ecology of chum salmon fry (*Oncorhynchus keta*) in a small estuary. Journal of the Fisheries Research Board of Canada 31: 83-92.

Masson, D. and Peña, A. 2009. Chlorophyll distribution in a temperate estuary: The Strait of Georgia and Juan de Fuca Strait. Estuarine, Coastal and Shelf Science, 82(1): 19-28.

Morris, S., Leaney, A.J., Bell, L.M. and Thompson, J.M. 1979. The Courtenay River estuary status of environmental knowledge to 1978. Fisheries and Environment Canada Special Estuary Series No. 8.

Northcote, T.G. 1997. Potamodromy in Salmonidae: living and moving in the fast lane. North American Journal of Fisheries Management 17: 1029-1045.

Okey, T.A., Alidina, H.M., Lo, V. and Jessen, S. 2014. Effects of climate change on Canada's Pacific marine ecosystems: a summary of scientific knowledge. Reviews in Fish Biology and Fisheries 24(2): 519-559.

Romanuk, T.N. and Levings, C.D. 2005. Stable isotope analysis of trophic position and terrestrial versus marine carbon sources for juvenile Pacific salmonids in nearshore marine habitats. Fisheries Management and Ecology 11: 113-121

Ryan, B. 2008. Preliminary analyses of coal refuse material from Vancouver Island. Geoscience Reports 2008, BC. Ministry of Energy, Mines and Petroleum Resources. pp. 99-118.

Simenstad, C.A. and Fresh, K.L. 1995. Influence of intertidal aquaculture on benthic communities in Pacific Northwest estuaries: scales of disturbance. Estuaries 18: 43-70.

Tryon, L.C. 2012. The Importance of Life History Diversity in Coho Salmon. Master of Science in Environment and Management Thesis. Royal Roads University, Victoria, British Columbia.

Vandermeulen, H., Jamieson, G. and Ouellette, M. 2006. Shellfish aquaculture and marine habitat sensitivity case studies. Canadian Science Advisory Secretariat Research Document 2006/036.

Water Survey of Canada. 1978. Hydrometric data for Rosewall Creek. http://www.wsc.ec.gc.ca/applications/H2O/HydromatD-eng.cfm

Chapter 6

Baynes Sound as an Important Bird Area

Ron Ydenberg

Professor, Centre for Wildlife Ecology, Simon Fraser University, Burnaby, BC

INTRODUCTION

Baynes Sound is part of an 'Important Bird Area' (hereafter IBA) named 'K'omoks' (Chapter 1 Figure 1), which includes the waters around Denman and Hornby Islands, Comox Bay, a stretch of the coast of Vancouver Island north of Cape Lazo, the town of Comox, and the Comox Valley. Baynes Sound itself lies between Denman and Vancouver Islands, stretching from Deep Bay at the southern end to the Seal Islets at the northern end. The K'omoks IBA was formed by the amalgamation of three separate IBAs (Comox Valley IBA, Baynes Sound IBA and Lambert Channel/Hornby Island Waters IBA). These places won IBA status because they are the winter home for large numbers of waterbirds, including the trumpeter swans of the Comox Valley.

IBA status has formalized what many in British Columbia already knew, and confers no legal status. Nonetheless, this recognition is important because IBAs are acknowledged as places of international significance for the conservation of birds and biodiversity (see http://www.ibacanada.ca/). The IBA Program is coordinated by the NGO BirdLife International, with Bird Studies Canada and Nature Canada as Canadian co-partners. The IBA Program is a science-based initiative to identify, conserve, and monitor a network of sites that provide essential habitat for Canada's bird populations. IBAs are identified using internationally agreed upon criteria that are standardized, quantitative, and scientifically defensible. They are discrete sites that support threatened birds, large groups of birds, or birds restricted by range or by habitat. IBAs range in size from very tiny patches of habitat to large tracts of land or water. They may encompass private and public land, and may overlap with sites that have other forms of legal protection.

The coastline of British Columbia is long and complex, with deep fjords extending more than 100 km inland, countless bays and estuaries, many large and thousands of small islands with their associated straits, sounds, and passages. The oceanography of these coastal waters is accordingly complicated and variable. There are many accounts in the literature of places where birds congregate, but systematic surveys were not made until the 1970s. In January of 1977 and again in March of 1978, the Canadian Wildlife Service surveyed the coast to estimate the numbers of birds. Using techniques and sampling expertise developed over many years of practise on smaller scales (Vermeer et al. 1983), they flew thousands of kilometers at low altitude and counted birds on the water. The effort was stratified into three major and fifteen sub-regions, so that the entirety of British Columbia's coast was represented.

The results of these surveys relevant to this chapter are summarized in Table 1. Here the data reported by Vermeer et al. (1983, their Table 2) are divided into the sub-region 'Strait of Georgia along the east coast of Vancouver Island' (which includes Baynes Sound) and 'elsewhere' (all other survey areas). Note first of all that the effort was substantial, with a total of 12,642 km flown. A second point is that the estimates are reasonably consistent. No estimate of the error around these estimates is included in the report, but there is every reason to believe that the density estimates are sound.

Table 1   The density of birds (individuals per survey kilometer) along the Pacific coast of Canada, as surveyed from the air by the Canadian Wildlife Service. These data are from Vermeer et al. (1983, their Table 2). For presentation here, the results are divided into 'East coast, Vancouver Island' (corresponding to Region A in Vermeer et al. 1983, Nanaimo to Campbell River, which includes Baynes Sound) and 'elsewhere' (all other survey areas). The 'elsewhere' totals were adjusted to exclude Baynes Sound. The survey counted all birds observed on the water, classified into major taxonomic groups (see Tables 2 and 3)

Region	Date	km Surveyed	Birds/km
East coast, Vancouver Island	Jan/Feb 1977	250	170
	March 1978	167	255
Elsewhere on BC coast	Jan/Feb 1977	6888	27
	March 1978	5337	40

Estimating the total number of waterbirds from these data requires an estimate of the total length of BC's coastline. Wikipedia reports a length of 27,725 km, and taking this value leads to a winter population estimate in excess of 600,000 waterbirds (see below for a list of taxa included). But this is uncertain. Due to the highly convoluted nature of BCs coast, estimating its length is not straightforward, and it is not clear that a kilometre of survey flight distance matches a kilometre of coastline as reported by Wikipedia. Moreover, the numbers in Table 1 do not include birds that were out of sight (i.e. diving below the surface, further offshore, or feeding onshore), so the total of 600,000 waterbirds is likely an underestimate, perhaps vastly so. But the salient point here is that the density of waterbirds along eastern shoreline of Vancouver Island is about six times that of the coast in general. In fact, the density is the highest of any of the 15 sub-regions included in the survey, and are rivalled in winter only by the Fraser River estuary, also an IBA, and internationally renowned for its winter and migratory bird populations. As we will see, the density in Baynes Sound is even higher than the general east coast of Vancouver Island estimate of about 170–250 birds per kilometer. Baynes Sound is unquestionably an important site for birds.

Table 2 The density (individuals per survey kilometer) in major taxonomic categories along the Pacific coast of Canada, as surveyed from the air by the Canadian Wildlife Service. The data are from Vermeer et al. (1983). For presentation here, the results are divided into 'East coast, Vancouver Island' (corresponding to Region A in Vermeer et al. 1983, Nanaimo to Campbell River, which includes Baynes Sound) and 'elsewhere' (all other survey areas). The 'elsewhere' totals were adjusted to exclude Baynes Sound. 'nr' = not reported

Group	East coast, Vancouver Island		Elsewhere	
	Jan/Feb 1977	*March 1978*	*Jan/Feb 1977*	*March 1978*
Loons	0.53	0.78	0.15	0.16
Grebes	6.30	19.40	0.79	1.44
Cormorants	0.58	1.20	0.46	1.72
Diving ducks	53.60	136.50	11.30	23.69
Dabbling ducks	58.50	7.70	7.61	1.92
Swans & Geese	1.00	20.30	0.83	1.37
Gulls	50.00	69.30	5.24	9.05
Auks	0.00	0.05	0.36	0.04
Unidentified	nr	nr	nr	nr

Vermeer et al. (1983) assigned the birds seen into several major taxonomic categories, reported here in Tables 2 and 3. As in Table 1 we have divided their data into the sub-region 'Strait of Georgia along the east coast of Vancouver Island' (which includes Baynes Sound) and 'elsewhere' (all other survey areas). The highest densities in Baynes Sound are attributable to the ducks (> 100 per kilometer) and the gulls

(> 50 per kilometer). As already noted in Table 1 the density of most taxa is several fold higher along the east coast of Vancouver Island than on the coast in general, with the exceptions of the categories 'cormorants' and 'auks'. Among the ducks, the three species of scoters (indistinguishable from the air) represent the single largest category in the Vermeer et al. (1983), reaching densities of 30-60 birds per kilometer. Thus, Baynes Sound is not only home for the largest densities of waterbirds along the entire BC coast, but it is home for many different species, and for some of these it is home to a large wintering concentration.

We will see below that the density of scoters in Baynes Sounds currently appears to be higher than 35 years ago, when the data presented in Table 1 were gathered. What is the status of other bird species? Crewe et al. (2012) summarize data from the British Columbia Coastal Waterbird Survey, a citizen science long-term monitoring program implemented by Bird Studies Canada to assess population trends and ecological needs of waterbirds using the province's coastal and inshore marine habitats. They analyzed standard monthly counts from many sites within the Strait of Georgia over a 12 year period (1999–2010) to assess trends in 57 waterbird species. They report that 33 species show stable populations or no trend, 22 species showed significantly declining trends, and just three species showed significant increasing trends. Among those that showed a declining trend were Western and Horned Grebes, Common, Red-throated and Pacific Loons, and several sea ducks (Black and White-winged Scoters, Long-tailed Duck, Barrow's Goldeneye, Harlequin Duck). The Vermeer et al. (1983) data indicate that Baynes Sound is very important for several of these declining groups. In the stable/no trend group were 20 species for which the Salish Sea is of recognised global or continental conservation importance.

ECOLOGY OF BAYNES SOUND RELEVANT TO WINTERING WATERBIRDS

Figure 1 illustrates an important aspect of the avian ecology of Baynes Sound, that moreover applies to the entire Strait of Georgia, and indeed to much of the BC coast: Compared with winter the diversity of birds is low during summer. The immediate reason is that most of the loons, grebes, swans, geese, and ducks depart for distant, inland breeding areas. Other groups such as the gulls and cormorants leave to congregate at breeding colonies in and around the Strait of Georgia. Ducks of 21 species represent about 60% of the total number of individuals in winter, and so not only the diversity but also the density of birds is low during summer, an observation that also applies to many other coastal waters of BC.

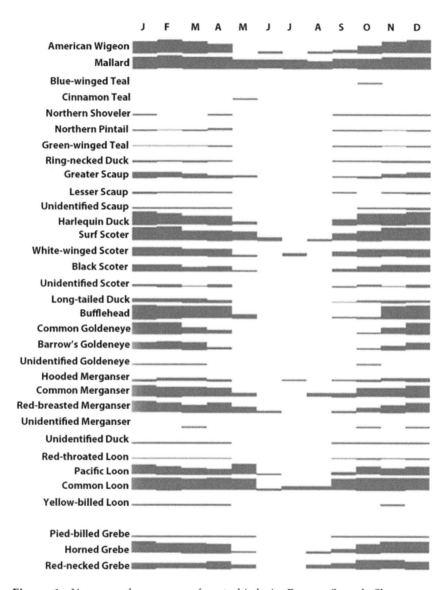

Figure 1 Year-round presence of waterbirds in Baynes Sound. Shown are monthly indices of frequency (mean number of surveys where the species was detected) for all species seen at Buckley Bay, in the middle of Baynes Sound, from British Columbia Coastal Waterbird Survey data (see http://www.bsc-eoc.org/volunteer/bccws/index.jsp?targetpg=bccwsdata&lang=EN). Many species are absent April – August.

There are several reasons for the high winter density of birds along the eastern coast of Vancouver Island, and each of these factors applies with extra force to Baynes Sound. First, winters are generally mild, and Georgia Strait is largely sheltered by the mountainous spine of Vancouver Island from Pacific winter storms, which are predominantly southwesterly. This applies especially to its western shore, including Baynes Sound. On its eastern side Baynes Sound is protected by Denman Island from the strong, cold Arctic outflows that periodically blow out of the deep fjords along the mainland coast.

The orientation of Baynes Sound along the north-south tidal axis of the Strait of Georgia and the openings at its north and south ends give it excellent tidal flushing and hence large amounts of food are swept through with both rising (northward flowing) and falling (southward flowing) tides. The Strait of Georgia is highly productive (see Vermeer and Butler 1989), and so the large tidal exchange delivers a large and regular influx of food to feed the zooplankton, benthos and fish in Baynes Sound, which in turn feed the birds. Many of the shellfish, other benthic creatures and fish reliant on this influx are thus able to be active and grow even during the winter (e.g. Lacroix 2001; Zydelis et al. 2006). Baynes Sound further has a diversity of aquatic habitats, including extensive muddy, sandy as well as rocky intertidal areas. These diversify the producer (i.e. prey) base, and hence the diversity of consumers (i.e. birds). Abundant and well-distributed freshwater inputs make it additionally attractive. And during winter, high tides in the Strait of Georgia occur mostly during daylight hours, which is advantageous for visual foragers that obtain their food by diving in the intertidal region, such as most of the diving ducks. A final reason for the attractiveness of Baynes Sound is that a major herring spawning event (Chapter 4) takes place around the spring equinox, and is visited by tens of thousands of birds. This event probably accounts for the higher tallies recorded during March (see Table 1).

SCOTERS AND SHELLFISH

Among waterbirds, diving ducks are the most abundant group along the entire British Columbia coastline (Table 2), a fact that holds even more strongly along the east coast of Vancouver Island. And among the diving ducks, scoters are the single largest group (Table 3). The three scoter species are combined in Tables 2 and 3, as they are indistinguishable from the air. In order of abundance they are the surf scoter (*Melanitta perspicillata*), the white-winged scoter (*M. fuscus*), and the black scoter (*M. nigra*). Numbers of the latter are far smaller, and represent perhaps a few per cent of the total, while surf scoters represent the great majority (perhaps 80%).

Table 3 The density (individuals per survey kilometer) of duck species along the Pacific coast of Canada, as surveyed from the air by the Canadian Wildlife Service. The data are from Vermeer et al. (1983). For presentation here, the results are divided into 'East coast, Vancouver Island' (corresponding to Region A in Vermeer et al. 1983, Nanaimo to Campbell River, which includes Baynes Sound) and 'elsewhere' (all other survey areas). The 'elsewhere' totals were adjusted to exclude Baynes Sound. 'nr' = not reported

Group	East coast, Vancouver Island		Elsewhere	
	Jan/Feb 1977	*March 1978*	*Jan/Feb 1977*	*March 1978*
Scoters	33.80	58.50	3.63	16.40
Goldeneyes	4.00	12.10	4.44	2.33
Scaups	9.10	24.60	0.86	1.80
Buffleheads	2.10	4.40	0.61	0.93
Long-tailed ducks	0.04	17.10	0.10	1.04
Harlequins	0.25	0.96	0.02	0.17
Mergansers	0.90	18.50	0.56	0.99
Unid.diving ducks	3.40	0.29	1.09	0.09
Mallards	27.90	2.00	2.51	0.45
Wigeons	25.10	5.70	2.62	0.09
Unid. dabblers	5.60	nr	2.47	nr

Baynes Sound attracts so many scoters for the same reason it attracted human shellfish gatherers since humans first settled there, perhaps 9,000 years ago. Its broad intertidal areas, sheltered waters, mild conditions and high tidal exchange offer perfect conditions for large numbers of bivalves, especially clams. Though they consume a wide variety of marine invertebrates during the winter as well as herring spawn when available, scoters are specialized for the consumption of bivalves, which they obtain by diving. Surf scoters focus on mussels which they wrench from the surfaces on which they grow, while white-winged scoters eat clams, excavated from the sediment. Small items may be ingested under water but larger items are retrieved to the surface. All are swallowed whole, crushed in the muscular gizzard, and passed through the gut rapidly (~30 minutes). Shell fragments are excreted. Birds often feed in large flocks, and because the intake rate of their large and bulky prey exceeds the processing rate of the digestive system, scoters feed in bouts of intense diving alternating with periods of rest during which they move further offshore.

The numbers of surf and white-winged scoters in Baynes Sound were counted in 1981 and again during the winters of 2002-2005. These data are presented in Table 4, which shows that the numbers of both scoter species have tripled over this period. These increases are undoubtedly related to the increased availability of food, because they are largest in Mud Bay and Fanny Bay (where shellfish aquaculture increased clam densities) and in Comox Harbor (where the invasive varnish clam greatly increased the numbers of clams). In accord with their diets, white-winged scoters outnumber surf scoters in Baynes Sound, even though they constitute perhaps 20% of the scoters on the coast as a whole.

Table 4   The total numbers of surf and white-winged scoters counted in Baynes Sound on ground surveys. The 1981 data are from Dawe et al. (1998). The 2002-05 data come from the shore-based surveys carried out by Deb Lacroix and summarized/analysed by Ramunas Zydelis (D. Esler, pers. comm.) and are as yet unpublished. These surveys were carefully matched to the exact locations surveyed by Dawe et al. (1998). Both species have increased about three-fold

Region	Surf scoters			White-winged scoters		
	1981	*2002-5*	*Ratio*	*1981*	*2002-5*	*Ratio*
Deep Bay	128	197	1.54	244	449	1.84
Mud Bay	104	343	3.30	202	253	1.25
Fanny Bay	142	473	3.33	367	996	2.71
Central	99	107	1.08	185	37	0.20
North	83	160	1.93	158	360	2.28
Comox	364	1469	4.04	221	1691	7.65
Total	**920**	**2749**	**2.99**	**1377**	**3786**	**2.75**

That such a population of scoters makes a measurable impact on the density of clams in the course of a winter is shown in Figure 2. The proportional reduction matches that reported by Bendell and Ydenberg (2001), based on measurements made in the mid-intertidal of Sandy Island Provincial Park (i.e. no aquaculture or commercial harvesting), though the density at Sandy Island is about twice that reported in Figure 2. Bendell and Ydenberg (2001) also show that the removal of clams is size-dependent, presumably because scoters preferentially take larger clams. The shellfish industry protects extensive portions of their leases with nets that limit the access of scoters and presumably also large excavating crabs. Small clams of several species are able to move between sites by leaving their burrows and 'swash-riding' (Ellers 1995) or by secreting a mucous parachute to move in currents. Many small individual clams likely move from sites under nets to escape competition from larger clams, and thus continually replenishing sites accessible to scoters, where they are continually removed. A better understanding of such trophic interactions would help evaluate the economic and environmental costs and benefits of nets.

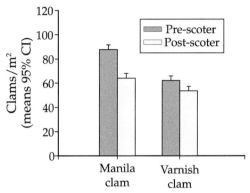

Figure 2 The impact of scoters on the density of manila and varnish clams in Baynes Sound, measured prior to scoter migratory arrival in October, and after migratory departure in April. Summarized from Lewis et al. (2007).

In summary, Baynes Sound's status an Important Bird Area is well-deserved. It houses a high density of wintering waterbirds, rivaled on the entire BC coast only by the Fraser estuary. The diversity of waterbirds is also great.

REFERENCES

Bendell-Young L.I. and Ydenberg R.C. 2001. Ecological Implications of the Shellfishery; A Case Study on the West Coast of British Columbia, Canada. pp. 57-70. *In*: Waters in Peril, Bendell-Young L., Gallaugher P. (eds) Springer, Boston, MA, doi:10.1007/978-1-4615-1493-0.

Crewe, T., Barry, K., Davidson, P. and Lepage, D. 2012. Coastal waterbird population trends in the Strait of Georgia 1999–2011: Results from the first 12 1983s of the British Columbia Coastal Waterbird Survey. British Columbia Birds 22: 8-35.

Dawe, N.K., Buechert, R. and Trethewey, D.E.C. 1998. Bird use of Baynes Sound - Comox Harbour, Vancouver Island, British Columbia. 1980-1981. Technical Report Series No. 286, Canadian Wildlife Service, Pacific and Yukon Region, British Columbia. 198 pp.

Ellers, O. 1995. Behavioral Control of Swash-Riding in the Clam *Donax variabilis*. The Biological Bulletin 189: 120-127.

Lacroix, D.L. 2001. Foraging impacts and patterns of wintering surf scoters feeding on bay mussels in coastal Strait of Georgia, British Columbia. M.Sc. thesis, Simon Fraser University.

Lewis, T.L., Esler, D. and Boyd, W.S. 2007. Effects of predation by sea ducks on clam abundance in soft-bottom intertidal habitats. Marine Ecology Progress Series 329: 131-144.

Vermeer, K., Campbell, R.W., Robertson, I., Kaiser, G. and Lemon, N. 1983. Distribution and densities of marine birds on the canadian west coast. Special Publication. Canadian Wildlife Service, Ottawa. 80 p.

Vermeer, K. and Butler, R.W. (eds). 1989. The Ecology and Status of Marine and Shoreline Birds in the Strait of Georgia, British Columbia. Special Publication, Canadian Wildlife Service, Ottawa. 186 p.

Zydelis R., Esler, D., Boyd, S., LaCroix, D. and Kirk, M. 2006. Habitat use by wintering surf and white-winged scoters: effects of environmental attributes and shellfish aquaculture. Journal of Wildlife Management 70: 1754-1762.

Chapter 7

Seaweed Harvesting: A controversial new industry on the east coast of Vancouver Island, British Columbia

Ian K Birtwell

Retired Scientist, Fisheries and Oceans Canada, Bowser, BC

INTRODUCTION

A "pilot" seaweed harvest was authorized by the government of British Columbia in 2007. It was focussed on beaches close to the small semi-rural communities of Deep Bay and Bowser at the south-eastern entrance to Baynes Sound. These marine shores contain valuable habitat supporting populations of terrestrial and aquatic organisms, as well as Aboriginal, recreational and commercial fisheries. They provide food, spawning habitats, nursery and rearing habitats, riparian habitats, and migration pathways for many species of fish, birds and mammals. The adjacent area of Baynes Sound continues to be subjected to a continually expanding shellfish industry that supplies approximately 50 percent of British Columbia's total shellfish aquaculture production. Although the Sound and contiguous waters are economically and ecologically valuable a comprehensive management plan for the region has not yet emerged. Such a plan would build on assessments of both the economic assets and ecological disturbances that may be influencing important ecosystem processes.

Jamieson et al. (2001) provided a comprehensive ecological review of the impacts of inter-tidal shellfish aquaculture industry in Baynes Sound, but since seaweed harvesting was not occurring at that time the potential impacts of this additional industry were not addressed. Also, the timing and occurrence of spawning by important forage fish (e.g., surf smelt and Pacific sand lance) in Baynes Sound had not been investigated.

The scientific literature on the ecology of near shore environments is extensive and identifies the significant role of seaweeds within coastal food webs. Harvesting of seaweed has occurred on the east coast of Canada for more than a century; however the "Irish Moss" industry collapsed in 2012 on Prince Edward Island, and in other locations harvest rates could not be sustained over many years resulting in impacts to local economies and marine habitats (Anderson 2006; Chopin and Ugarte 2007; Vandermeulen 2013). Accordingly seaweed removal can be problematic and potentially become another constraint and concern to sustaining the integrity of aquatic communities, and, therefore, to the productivity in Baynes Sound and adjacent coastal waters. This concern, and those of people directly impacted by the seaweed harvest, have been expressed to federal and provincial government regulators and responsible agencies, local government, and politicians by the public and numerous organisations including Habitat, Salmonid Enhancement, and Streamkeepers Societies, BC Environmental Network, Sea Change Conservation Working Group, BC Shore Spawners Alliance, Association of Denman Island Marine Stewards, Islands Trust, retired scientists, and other individuals.

This chapter focuses on the harvesting of a species of red algae (*Mazzaella japonica*) in Baynes Sound, the ecological importance of seaweeds to coastal near shore areas, public and scientific concerns about future seaweed harvesting as well as legislative considerations, and public consultations.

LEGISLATION, AND REGULATION OF SEAWEED HARVESTING IN BRITISH COLUMBIA

Aspects of both provincial and federal legislation relate to the seaweed harvest which may occur between September and February (before herring spawning).

The BC Ministry of Agriculture is responsible for the management of the commercial harvest of marine plants in British Columbia. Their mandate is to:

"ensure that the harvest of marine plants is done in an approved manner, and that the harvest will not compromise habitat or traditional First Nations use of the resource". The responsible *"minister may suspend, revoke or refuse to issue a licence if* "(a) the licensee has failed to comply with a condition of a licence, or (b) the harvesting of kelp or other aquatic plants under the licence would i) *tend to impair or destroy a bed or part of a bed on which kelp or other aquatic plants grow*, ii) *tend to impair or destroy the supply of any food for fish*, or iii) *be detrimental to fish life."* (Birtwell et al. 2013).

The context of this aspect of regulation relates to the federal *Fisheries Act* (Government of Canada 2012; R.S.C. 1985, c. F-14, Last amended on

April 5, 2016). The term "fish" includes (*a*) parts of fish, (*b*) shellfish, crustaceans, marine animals and any parts of shellfish, crustaceans or marine animals, and (*c*) the eggs, sperm, spawn, larvae, spat and juvenile stages of fish, shellfish, crustaceans and marine animals, and "fish habitat" - the spawning grounds and any other areas, including nursery, rearing, food supply and migration areas on which fish depend directly or indirectly in order to carry out their life processes.

A democratically-derived Official Community Plan (OCP), within the administrative boundaries of the Regional District of Nanaimo on Vancouver Island, includes guidelines for the management of shoreline areas (Regional District of Nanaimo 2017). Pertinent aspects of the OCP stipulate that: the shore zone must be recognized as a finite resource; policies should support enhancement of the physical, recreational, and visual qualities of the area, while recognizing the relationship of upland foreshore land uses; the area along the entire coastal waterfront area within 1000 m from the foreshore is recognized as environmentally sensitive - from the Deep Bay Spit to the eastern boundary of Area 'H' and, that development which would alienate the foreshore from public access or impact on the natural environment is discouraged. The specific "permitted" area for seaweed harvesting has been on beaches covered by the OCP. These areas are recognized as non-industrial (c.f. Baynes Sound proper), where vehicular traffic is locally prohibited, and where access to the marine foreshore is limited.

An independent review of legislative responsibilities relating to the harvesting of seaweed was undertaken by the Environmental Law Clinic and the University of Victoria (Brusse et al. 2013). They concluded the following:

- The ongoing provincial licensing program appears to encourage non-compliance with the federal *Fisheries Act*, by allowing harvesters to damage fish habitat which provides vital sources of food for larger fisheries;
- Contrary to its own policies, the Ministry of Agriculture has been licensing and encouraging a new commercial seaweed industry without adequate study of habitat impacts from the harvest.
- DFO has been unresponsive to requests that it investigates potential non-compliance with section 35 of the federal *Fisheries Act* and, has failed to take proactive steps to set proactive and protective conditions on the harvest.
- The two levels of government involved are abdicating responsible oversight over a new industry with potentially serious environmental impacts. They appear to be unable or unwilling to coordinate proper oversight.
- The provincial minister should exercise discretion to revoke the licenses, until: further studies are completed; and DFO is able

to adequately address any non-compliance occurring under its jurisdiction – either through enforcement of section 35(1) or by issuing proactive section 35(2) authorization orders with stringent protective conditions.

COMMERCIAL VALUE OF SEAWEEDS

The world-wide commercial interest in seaweeds relates to their use in fertilizers, animal foods, and especially to the value of their phycocolloids (the major polysaccharides: alginates, carrageenans, agars, fucanes, laminarans, ulvans, and floridean starch). The annual global production of phycocolloids is about 100,000 tonnes, with a gross market value of $1 billion (US); 80% of the global agar and carrageenan production and 30% of the global alginate production is used in the food industry for their gelling, thickening and stabilizing properties (refer to Jaspers and Folmar 2013). In 1995 annual carrageenan sales were over $200 million (US) or about 15% of the world use of hydrocolloids (Bixler 1996).

ECOLOGICAL VALUE OF SEAWEEDS

Seaweeds are, fundamentally, of high ecological importance (Harley et al. 2012) and accordingly their removal, whether while living, or dead, will have an ecological impact. The scale of the impact depends on the location and nature of the harvest, its timing, the methods used to harvest, and the impact on organisms and their role in the ecosystem processes and productivity. Seaweeds are essential valued ecosystem components that sustain other aquatic organisms, including those that support economically-important commercial, recreational, and Aboriginal finfish and shellfish fisheries. They also contribute to the maintenance of primary and secondary productivity. Seaweeds are, therefore, a basic and vital component of the marine ecosystem of significant importance to the protection of fish and their habitat as required under the *Fisheries Act* (Government of Canada 2012).

Living, dead and decomposing algae provide food for many components of food webs. Aside from the physical aspects of algae and the role they play in the structural complexity of waters which constitute fish habitat, this "primary production" has a direct influence on those organisms higher in the food chain (Levings et al. 1983). Algae and other plant material are part of this primary production which provides nourishment while alive but also when dead and decaying and producing detritus. Mann (1988) stated that "more energy and materials flow through detrital food webs than through grazer food webs", meaning that more energy is transmitted to other trophic levels from dead decomposing plant tissue than from living tissue consumed by grazers. In addition, the nature of primary production influences how detritus may be beneficial to other organisms (Jones and Iwama 1991; Rodhouse and Roden 1987). For example, marsh grasses and other vascular plants require a longer time to be broken down by fungi and bacteria into smaller particles than do algae which decompose at faster rates and are more nutritious and available (Mann 1988).

Animals can obtain much nourishment directly from algal material (Findlay and Tenore 1982; Tenore 1981 and 1988). Thus the importance of particulate macroalgae detritus is documented and emphasized due to its significant and important role in the productivity of invertebrates (e.g. crustaceans, Rossi et al. 2010; bivalve molluscs, Crosby et al. 1989; gastropods:, Smith et al. 1985; fish, Levings et al. 1983; Mann 1988). Some inter-tidal fish have been reported to directly use algae as food, such as the cockscomb pricklebacks (Peppar 1965) which occur in the lagoons at Bowser/Deep Bay. Benthic inter-tidal communities, especially crustaceans, within Georgia Strait have been reported to be important in the diet of many juvenile fish such as salmon (Levings et al. 1983). Furthermore, the net shoreline movement of materials floating in sea water is into Baynes Sound proper from the Deep Bay/Bowser area. Thus algae and their breakdown products and associated organisms are transported into waters upon which wild and cultured shellfish and other valuable organisms derive nourishment.

Seaweed that washes ashore and becomes stranded is termed 'wrack'. It is a complex mixture of vegetated materials from vascular plants and seaweeds, and dead and dying organisms with an associated community of micro and macro organisms. In some locations in the Deep Bay/ Bowser area seaweed that is washed up along shorelines accumulates in dense mats often exceeding a metre deep (personal observation). Wave and tidal actions continuously move this material and large quantities become mixed into the beach substrate. During tidal changes it can be re-suspended, fractured, decomposed and transported with subsequent wave action.

Carefoot (2014) summarized the nourishing role of wrack along shorelines. He noted that the particles and pieces of seaweed find their way back into the intertidal and sub tidal regions where they are consumed by various suspension-feeding and deposit-feeding crustaceans, worms, and sea urchins, and many omnivorous and herbivorous invertebrates. Clearly wrack plays a vital role as a conduit of energy and nutrients from the ocean to the land and back again (Carefoot 2014). Therefore, the onshore deposition of macroalgae and macrophyte wrack provides a potentially significant marine subsidy to inter-tidal and supra-tidal herbivore and decomposer communities. Based on the study of daily input loads to beaches, Orr et al. (2005) estimated the summer wrack deposition in Barkley Sound, British Columbia where cobble beaches retained approximately 10 times and 30 times more wrack than gravel and sand beaches, respectively. The beaches upon and in which detached seaweed occurs in the area of Deep Bay and Bowser are primarily of cobble, pebble and coarse sand.

Tyron (2012) also emphasized the ecological importance of wrack in supporting a diversity of animals. She commented on the association of wrack with highly diverse infaunal assemblages and their predators, including taltrid amphipods and staphylid beetles (Richards 1984), oligochaetes and nematodes (Sobocinski 2003) and birds (Bradley and Bradley 1993). Rossi et al. (2010) documented the importance of seaweed wrack derived from an invasive species of algae (*Sargassum muticum*), present in the inter-tidal areas of Deep Bay and Bowser, to sustain part of the benthic food web (it is also a significant substrate for herring egg attachment). Similarly, McGwynne et al. (1988) and Olabarria et al. (2010) commented on the relationship of buried and decaying seaweed wrack to beach organisms and the role it plays in influencing the composition and structure of meiofaunal and macrofaunal assemblages, respectively. Lastra et al. (2008) also reported the importance of beach cast materials to invertebrate populations and community structure in the inter-tidal zone, and Pennings et al. (2000) commented that invertebrate consumers (an isopod, and rocky and sandy-shore amphipods) tended to prefer wrack over fresh seaweeds of the same species. Romanuk and Levings (2003) documented the increased importance of such vegetated material to organisms that dwell in the transitional supra-littoral zone between the terrestrial and aquatic environment.

The diets of commercially important fish such as juvenile salmonids, herring and surf smelt overlap with the invertebrate prey in beach wrack. For example, epibenthic crustaceans, including amphipods, provide important food web connections for salmon in Puget Sound (Brennan and Culverwell 2004) and in British Columbia (Levings and Jamieson 2001; Romanuk and Levings 2005). This trophic relationship has important implications for fisheries values and salmonids and forage fish, along with many other species, consume amphipods, worms, and insects that become dislodged and available in the inter-tidal zone upon tidal immersion. Many birds also nest and feed on beach wrack communities (Bradley and Bradley 1993).

Wrack is an important contributor of nitrogen and carbon cycling in marine and terrestrial environments due to the relatively rapid release of nutrients during breakdown, which facilitates primary productivity (benthic algae and phytoplankton) and benefits higher trophic levels in the food chain (Zempke-White et al. 2005; Mews et al. 2006). Both nutrients and carbon can be transported to sub-tidal zones (Romanuk and Levings 2006), to interstitial spaces in beaches (Dugan et al. 2011), and to marine riparian systems (Levings and Jamieson 2001; Polis and Hurd 1996). Furthermore, beach wrack accumulation on sandy beaches provides thermal insulation from temperature extremes and maintains a humid environment for the organisms that thrive in it and in the substrate below (Columbini and Chelazzi 2003; Tyron 2012).

Seaweeds are of functional significance even when they become detached and they may form floating masses which is habitat for numerous invertebrates and fish. They may also play a role in the dispersal of invertebrate species and juvenile fish (Zempke-White et al. 2005). Furthermore, when the seaweed and its inhabitants and organic matter wash back into the sea it can become habitat for juvenile fishes and a food source for herbivores and when further decomposed for filter feeders, the nutrients can be utilized by primary producers (Shaffer et al. 1995; Zempke-White et al. 2005).

Marine riparian zones and unconsolidated-sediment beaches provide critical habitat for marine fishes and invertebrates (Levings and Jamieson 2001). Marine riparian vegetation produces terrestrial insects, vital prey for foraging juvenile chinook salmon (Brennan and Culverwell 2004). Sandy/gravel beaches are spawning habitat for surf smelt and Pacific sand lance, forage fish species that are critical prey for hundreds of marine predators (Penttila 2007).

Inter-tidal areas with much habitat complexity occur within 3 km of Deep Bay. They support numerous organisms of high economic and ecological value especially in the unique and extensive tidal lagoons which are, in part, a legacy from the Qualicum First Nation who modified the shoreline and constructed fish traps and clam gardens about 3,000 to 5,000 years ago (*pers. comm.*, M. Racalma, Qualicum First Nation). Many organisms derive food from the inter-tidal lagoons and the adjacent beaches where detached seaweeds accumulate and have been harvested. Depending on the time of year, large numbers of ducks and geese, seals, sea lions and otters, eagles and humans can be found harvesting prey reliant upon these areas (Jamieson et al. 2001).

Eagles feed along the Deep Bay/Bowser shores in the summer time and this annual concentration around the Bowser lagoons often exceeds the highest numbers recorded during this season on the BC coast (Elliott et al. 2003). Fifty percent of the eagle's diet consists of the Plainfin midshipman (*Porichthys notatus*) which migrates in May from very deep waters to spawn in the inter-tidal area. Under paternal care, the larval fish hatch from eggs which adhere to the rock ceiling of the excavated nest and remain there until mid-August, after which time they move to protective vegetated (seaweeds and eelgrass) nursery habitats in the proximal lagoons (Bass 1995; Sisneros et al. 2009).

The Baynes Sound/Deep Bay-Bowser area is important habitat for forage fish which are typically abundant and provide food for other piscivorous animals, especially other fishes, marine mammals and birds. In the vicinity of Baynes Sound forage fish would include species such as herring (*Clupea pallasi*), surf smelt (*Hypomesus pretiosus*) and sand lance (*Ammodytes hexapterus*). However, there are many other species of fish that are important prey in the near shore marine food web, including species of cottids, gunnels and pricklebacks that reside in shallow, near shore habitats (Birtwell et al. 2013).

Several species of forage fishes are of vital importance to key commercial fish species, especially salmonids, rockfish, and halibut. These include, but are not limited to, sand lance, juvenile herring, and surf smelt. These three species are of special interest and they all spawn in shallow sub-tidal or inter-tidal habitats. Two of the species, sand lance and surf smelt, spawn at the same times, and at the same locations where the "pilot" seaweed harvest has occurred (Birtwell et al. 2013; de Graaf 2012). Once hatched, the larvae of sand lance and surf smelt can be found in the sand/gravel beaches and then in the adjacent shoreline aquatic habitats. Therefore, these fish are directly at risk from human activities on the beach and the timing of seaweed harvesting on the east coast of Vancouver Island and its location directly conflicts with, and impacts, the spawning and development of these fish.

COMMERCIAL SEAWEED HARVEST AND IMPACTS

The harvesting of seaweed and its artificial production occurs worldwide and about 13 million tonnes (fresh weight) annually has been collected (Zempke-White et al. 2005). The majority of this seaweed was harvested live (naturally growing and cultured) but there was also harvesting of material cast onto beaches.

Multi-Species Effects of Seaweed Harvesting

Lorensten et al. (2010), described for the first time the ramifications of seaweed removal practices to higher members of coastal food chains in Norway. Their results strongly suggest that kelp harvesting affects fish abundance and diminishes coastal seabird foraging efficiency. Schmidt et al. (2011) stated that marine vegetation provides important habitat, nitrogen, and carbon storage services, yet the extent of these services depends on the foundation species and its architecture. Changes in canopy structure will therefore have profound effects on associated food webs and ecosystem services. Thus, as increasing human pressures on coastal ecosystems threaten the continued supply of essential functions and services, the protection of marine vegetated habitats should be a management priority (Schmidt et al. 2011).

In 2009 the European Union declared that the commercial harvest for macroalgae should not be done in any way so as to cause a significant impact on ecosystems (Stagnol et al. 2013). On the east coast of Canada seaweed harvesting, which has occurred for decades, was not sustainable for all species targeted because of the methods used to collect the seaweed while it was growing, when detached, and the escalating quantities taken over time (Chopin and Ugarte 2007). The ecological consequences of harvesting were often considered only in relation to algal communities and re-growth in support of a sustainable harvest (Sharp and Pringle 1990; Chopin and Ugarte 2006). Less attention has been given to the ecological effects on the communities impacted by the harvest (Black and Miller 1994; Rangeley 1994; Lorentsen et al. 2010). Concern over the sustainability of the commercial ventures to harvest seaweed and the potential effects on communities of aquatic organisms has prompted regulations to be formulated to control the timing and modes of harvest, species of algae taken and allowable quotas. In certain circumstances a moratorium has been placed on these activities so that studies may be undertaken to assess impacts (e.g. New Zealand; Zempke-White et al. 2005).

A Review of Impacts of Harvesting Beach-Cast Seaweeds

A thorough review of the beach harvesting of seaweed was carried out by Zempke-White et al. (2005). They identified a number of key research gaps related to the removal of beach-cast seaweeds from the coastal environment. Knowledge gaps include: quantitative data on the distribution of beach-cast seaweeds; the relationship between beach-cast seaweed and off-shore algal stands; residence time of the seaweed on the beach; the fate of seaweeds when not collected and the communities they support; the role of floating seaweeds, and; the effects of removals on the coastal ecosystem and fisheries resources.

SEAWEED HARVESTING ALONG THE EAST COAST OF VANCOUVER ISLAND

The red algae *Mazzaella japonica* (considered a recent introduction to Canada related to proximal aquaculture operations) occurs close to the low tide level and in shallow sub-tidal waters especially around Deep Bay and Bowser. When detached, it is a target species for the commercial harvest from beaches which often receive substantial accumulations of algae following storms and powerful wave action, usually in the fall and winter.

In 2012, 5000 tonnes of *M. japonica* were licensed for removal from beaches during the fall and winter. In 2013 the shoreline for harvesting seaweeds was reduced from 22 km to 5 km (Deep Bay/Bowser), and the quantity to 600 tonnes; for the 2014/2015 fall/winter period the amount was increased to 900 tonnes. In 2015/2016 the total was increased to 1,500 tonnes (*pers. comm.*, G. Caine, Provincial government to G.R. Peterson, June 2014; Holden et al. 2018). At peak biomass during the harvest along the designated 4.2 km shoreline, the latter authors reported 1,586 tonnes of "highly transient" beach-cast seaweed and that the harvest removed less than 16% each week.

The harvest of seaweed in Deep Bay and Bowser relies on the use of tracked vehicles operating on beaches, the gathering of the beach-cast algae by pitch forks and rakes, and placing the collected material into large bags and mesh cages. The filled bags are loaded onto vehicles on the beach and/or the adjacent land, then to larger vehicles on roads for transport to drying and processing facilities.

In December 2012 the physical impact of a tracked seaweed carrier vehicle on spawning deposits of Pacific sand lance was assessed while it operated for a brief time at two activity levels over spawning deposits (de Graf 2014).

Pacific sand lance embryos were crushed and/or deformed at even low-impact levels (4-16% of typical daily impact). The results of this small experiment verified that serious harm to fish and their habitat occurred. The study did not assess such activity over a full harvesting day or of related collection pressure of material (e.g. by rakes and forks) and transportation activities and there was no assessment of sub-lethal effects. To this extent the study provided evidence of significant harm without addressing the full range of likely impacts. Accordingly, the results do not represent the full range or extent of impacts that could likely occur and are not representative of the worst case scenario.

The designated seaweed harvesting areas between Deep Bay and Bowser are within the most important herring spawning locations in BC (Hay and McCarter 1999 and 2006). The seaweed harvesting stipulations preclude harvesting during herring spawning times but these areas are also used by larval and juvenile herring for rearing and other important fish species spawn each year along these shores (Hay and McCarter 1997).

There has been little research related to the specific effects of harvesting *M. japonica* at Deep Bay and Bowser. Recent research by Holden et al. (2016, 2018) found that removal of *M. japonica* had inconclusive effects on invertebrates, however, however, higher trophic level organisms (e.g. fish) and their various sensitive live history stages (e.g. eggs) were not considered. Thus, there has not been a determination of the comprehensive role of the seaweeds in the food chain of the local ecosystems, and their physical, biological and chemical function.

CONCERNS

The foregoing comments documented the economic and ecological value of seaweeds and the potential impacts of harvesting. It is suggested that the appropriateness of a new industry on the east coast of Vancouver Island that would remove seaweeds from locations that currently support diverse natural and cultured organisms, often with links to sustainable and economically-important fisheries, is questionable and not supported scientifically. In addition the imposition of this activity is contrary to the wishes of many local residents and is in contravention of the OCP and seemingly some provisions of the *Fisheries Act*. Whether algal species may be cultured for their constituents in land-based facilities is perhaps a viable question considering the concerns expressed over harvesting in the natural environment.

Concerns about seaweed harvesting on the east coast of Vancouver Island were expressed following a review and assessment of scientific literature (Birtwell et al. 2013). Meetings were held between the authors of the review and provincial and federal agency representatives but no commitment was made to revise procedures which permit the harvest despite the absence of specific impact data and study. The provincial government does not have a documented basis for the seaweed harvest and its potential impact, and the federal government's Department of Fisheries and Oceans has expressed little concern. Neither government has researched the sustainability of the harvest and its ecological impact. Furthermore, the provincial government's "regulatory decision maker" has dismissed the scientific review with comments that "it's just a biological review"; "they've reviewed the literature and haven't done any research or provided any scientific data to the area"; "at the end of the day it's the people who want to participate (licence holders) who will decide if it is viable or not" (C. Wu, Jan 23, 2014; news@pqbnews.com), and it's "only 0.004% of the 26,000 km BC coastline" (http://www.cbc.ca/ontheisland/2013/06/26/is-seaweed-bcs-newest-cash-crop/).

The concerns, both pro and con the harvest, were expressed at a public meeting on November 27, 2013 and included the following:
- no prior public consultation with federal/provincial government and First Nations;
- disruption of peaceful environment;
- increased physical destruction of beach and foreshore;
- reduction in buffering against wave action increasing shoreline erosion;
- negative impact on property values;
- disregard for democratically-formulated "official community plan" and public use prohibition of vehicles on beach;

- ecological impacts and not enough research;
- seaweed is invasive and should be harvested from beach as soon as possible;
- beach cast seaweed is a nuisance, useless, and smells;
- harvesting is only for a short time;
- impact of vehicles no different than human footprint (harvester comment);
- provides seasonal employment; the product is valuable globally.

Public and scientific concern continues to be elevated and efforts to engage the responsible provincial and federal agencies have been pursued.

CONCLUSIONS AND RECOMMENDATIONS

There is a scientific basis for concern about the implementation of a new industry that would harvest seaweed along the east coast of Vancouver Island, and perhaps other areas in British Columbia. There is also substantial public concern over this harvest and together with the scientific ecological concerns they implicate governments for their failure to engage the public and organizations in meaningful consultations, and for acting unilaterally. Specific concerns are related to the lack of assessments of the potential impacts of the industry, its socio-economic advantages and disadvantages, and the wisdom of its initiation. It is most advisable that such studies be undertaken to ensure that if an industry were to develop it would be based on sound science-based decisions with low socio-economic and ecological risks; notwithstanding the need to address the legitimate personal and other concerns of those directly impacted by the harvesting activities. Also, it is important to recognize the present ecological value of the putative harvesting areas and the contribution of such areas to sustain commercial, recreational and Aboriginal fisheries and aquaculture.

Therefore, a scientific and ecological review of the *Mazzaella japonica* fishery is required; equivalent to the reviews usually conducted through a Department of Fisheries and Oceans Canadian Science Advisory Secretariat evaluation and reporting process. A thorough evaluation of the effects of seaweed harvesting should be undertaken in relation to the requirements of the impacted areas to support continued aquaculture activities and their future growth, and maintain the supporting habitat for other highly valuable components of the local ecosystem. This is a prerequisite to assist appropriate, sensible and sound decision-making based on pertinent factual information. The recommendations of Jamieson et al. (2001) regarding Baynes Sound are endorsed and should be reviewed and reconsidered in light of this emerging industry:

establish a multi-agency initiative to identify existing and potential future impacts; develop a network of protected areas in Baynes Sound that includes sensitive habitats, key bird habitats and which exclude shellfish culture; identify potential adverse impacts from inter-tidal shellfish aquaculture and implement mitigation where appropriate; consider inter-tidal aquaculture as both an economic asset and an ecological disturbance, and investigate the overall carrying capacity of the Baynes Sound ecosystem with respect to phytoplankton production and its removal by filter feeders. Furthermore, restrictions should be specified to protect certain ecologically valuable areas from any future harvesting (e.g. the inter-tidal pool and lagoon area within 3 km of Deep Bay, unconsolidated-sediment areas comprising spawning beaches for forage fish, and marine riparian habitats).

It seems to this author only common sense that a moratorium on seaweed harvesting and licensing should be imposed until the socio-economic and ecological impacts of the *Mazzaella* "fishery" have been identified, thoroughly assessed, and discussed with all concerned parties.

REFERENCES

Anderson, S. 2006. How sustainable are emerging low-trophic level fisheries on the Scotian Shelf? B.Sc. Thesis, Dalhousie University, Environmental Science, Halifax, Nova Scotia, Canada. 68 p.

Bass, A.H. 1995. Alternative life history strategies and dimorphic males in an acoustic communication system. pp. 258-260. *In*: Goetz, F.W. and Thomas, P. (eds). Proceedings of the Fifth International Symposium on the Reproductive Physiology of Fish. Austin, Texas.

Birtwell, I.K., de Graaf, R.C., Hay, D.E. and Peterson, G.R. 2013. Seaweed harvesting on the east coast of Vancouver Island, BC: a biological review. 28p. http://nilecreek.org/wp-content/uploads/2013/08/SEWEED-HARVESTING-BIOLOGICAL-REVIEW.pdf

Bixler, H.J. 1996. Recent developments in manufacturing and marketing carrageenan. pp. 35-57. *In*: Lindstrom, S.C. and Chapman, D.J. (eds). Proceedings of the Fifteenth International Seaweed Symposium, Valdivia, Chile, January 1995. Developments in Hydrobiology, vol 116. Springer, Dordrecht.

Black, R. and Miller, R.J. 1994. The effects of seaweed harvesting on fishes: a response. Environmental Biology of Fishes. 39: 325-328.

Bradley, R.A. and Bradley, D.W. 1993. Wintering shorebirds increase after kelp (Macrocystis) recovery. The Condor. 95(2): 372-376.

Brennan, J.S. and Culverwell, H. 2004. Marine Riparian: An Assessment of Riparian Functions in Marine Ecosystems. Washington Sea Grant Program. U.W. Board of Regents. Seattle, WA. 34 p.

Brusse, P., Jackson, G. and Sandborn, C. 2013. Seaweed harvesting on vancouver island: A new industry that requires better regulation. Prepared for the

Mid Vancouver Island Habitat Enhancement Society and the Nile Creek Enhancement Society. Environmental Law Clinic, University of Victoria, BC. 38 p. http://www.elc.uvic.ca/publications/documents/2013-02-05-Seaweed HarvestReport.pdf.

Carefoot, T.C., 2014. http://biodiversitybc.blogspot.ca/search?updated-min=2013-01-01T00:00:00-08:00&updated-max=2014-01-01T00:00:00-08:00&max-results=50

Chopin, T. and Ugarte, R. 2006. The seaweed resources of eastern Canada. University of New Brunswick, Centre for Coastal Studies and Aquaculture. 46p.

Columbini, I. and Chelazzi, L. 2003. The influence of marine allothonous input on sandy beach communities. Oceanography and Marine Biology: Annual Review. 115–159.

Crosby, M.P., Langdon, C.J. and Newell, R.I.E. 1989. Importance of refractory plant material to the carbon budget of the oyster *Crassostrea virginica*. Mar. Biol. 100(3): 343-352.

de Graaf, R.C. 2014. Potential impacts of beach wrack seaweed harvesting to inter-tidal forage fish spawning habitat. Unpublished report Emerald Sea Biological.

Dugan, J.E., Hubbard, D.M., Page, H.M. and Schimel, J.P. 2011. Marine macrophyte wrack inputs and dissolved nutrients in beach sands. Estuaries and Coasts, 34(4): 839-850.

Elliott, K.H., Struik, C.L. and Elliott, J.E. 2003. Bald Eagles, *Haliaeetus leucocephalus*, feeding on Spawning Plainfin Midshipman, *Porichthys notatus*, at Crescent Beach, British Columbia. Canadian Field-Naturalist. 117(4): 601-604.

Findlay, S.E, and Tenore, K.R. 1982. Nitrogen source for a detritivore: detritus substrate versus associated microbes. Science. 218: 371-2.

Government of Canada 2012. Fisheries Act R.S.C., 1985, c. F-14, amended April 5, 2016, Published by the Minister of Justice, Ottawa Canada. 70p. http://laws-lois.justice.gc.ca.

Harley, C.D.G., Anderson, K.M., Demes, K.W., Jorve, J.P., Kordas, R.L., Coyle, T.A. and Graham, M.H. 2012. Effects of climate change on global seaweed communities. Journal of Phycology. 48(5): 1064-1078.

Hay, D.E. and McCarter, P.B. 1997. Larval retention and stock structure of British Columbia herring. J. Fish. Biol. 51: 155-175.

Hay, D.E. and McCarter, P.B. 1999. Distribution and timing of herring spawning in British Columbia. Canadian Stock Assessment Secretariat Research Document 99/14: (Working Paper H98-5).

Hay, D.E. and McCarter, P.B. 2006. Herring spawning areas of British Columbia: A review, geographical analysis and classification. Revised MS Rept. 2019. http://www.pac.dfo-mpo.gc.ca/science/species-especes/pelagic-pelagique/herring-hareng/herspawn/pages/project-eng.htm

Holden, J.J., Dudas, S. and Juanes, F. 2016. Is commercial harvesting of beach-cast seaweed ecologically sustainable? Integr Environ Assess Manag. 12: 825-827.

Holden, J.J., Kingzett, B.C., MacNeill, S., Smith, W., Juanes, F. and Dudas, S.E. 2018. Beach-cast biomass and commercial harvesting of a non-indigenous seaweed, *Mazzaella japonica*, on the east coast of Vancouver Island, British Columbia. Journal of Applied Phycology. 30: 1175-1184.

Jamieson, G.S., Chew, L., Gillespie, G., Robinson, A., Bendell-Young, L., Heath, W., Bravender, B., Nishimura, D. and Doucette, P. 2001. Phase 0 review of the environmental impacts of inter-tidal shellfish aquaculture in Baynes Sound. Fisheries and Oceans Canada, Canadian Science Advisory Secretariat, Ottawa, Ontario. 103 p.

Jaspers, M. and Folmar, F. 2013. Sea Vegetables for Health. A report prepared for Food and Health Innovation Service. Department of Chemistry, School of Natural and Computing Science, University of Aberdeen, Scotland. 29 p.

Jones, T.O. and Iwama, G.K. 1991. Polyculture of the Pacific oyster, *Crassostrea gigas* (Thunberg), with chinook salmon, *Oncorhynchus tshawytscha*. Aquaculture. 92: 313-322.

Lastra, M., Page, H.M., Dugan, J.E., Hubbard, D.M. and Rodil, I.F. 2008. Processing of allochthonous macrophyte subsidies by sandy beach consumers: estimates of feeding rates and impacts on food resources. Marine Biology. 154: 163-174.

Levings, C.D., Foreman, R.E. and Tunnicliffe, V.J. 1983. Review of the benthos of the Strait of Georgia and Contiguous Fjords. Canadian Journal of Fisheries and Aquatic Sciences 40: 1120-1141.

Levings, C.D. and Jamieson, G. 2001. Marine and estuarine riparian habitats and their role in coastal ecosystems, pacific region. Fisheries and Oceans Canada, Canadian Science Advisory Secretariat, Ottawa, Ontario. 42 p.

Lorentsen, S.-H., Sjøtun, K. and Grémillet, D. 2010. Multi-trophic consequences of kelp harvest. Biological Conservation. 143(9): 2054-2062.

Mann, K.H. 1988. Production and use of detritus in various freshwater, estuarine, and coastal marine ecosystems. Limnol. Oceanogr. 33: 910-930.

McGwynne, L.E., McLachlan, A. and Furstenberg, J.P. 1988. Wrack breakdown on sandy beaches-its impact on interstitial Meiofauna. Marine Environmental Research. 25: 213-232.

Mews, M., Zimmer, M. and Jelinski, D.E. 2006. Species-specific decomposition rates of beach-cast wrack in barkley sound, British Columbia, Canada. Marine Ecology Progress Series. 328: 155-160.

Olabarria, C., Incera, M., Garrido, J. and Rossi, F. 2010. The effect of wrack composition and diversity on macrofaunal assemblages in inter-tidal marine sediments. Journal of Experimental Marine Biology and Ecology. 396(1): 18–26.

Orr, M., Zimmer, M., Jelinski, D.E. and Mews, M. 2005. Wrack deposition on different beach types: spatial and temporal variation in the pattern of subsidy. Ecology. 86: 1496-1507.

Pennings, S.C., Carefoot, T.H., Zimmer, M., Danko, J.P. and Ziegler, A. 2000. Feeding preferences of supra-littoral isopods and amphipods. Canadian Journal of Zoology. 78(11): 1918-1929.

Penttila, D. 2007. Marine forage fishes of puget sound. Puget Sound Near shore Partnership report No. 2007-03. Published by Seattle District, U.W. Army Corps of Engineers, Seattle, Washington.

Peppar, J.L. 1965. Some features of the life history of the cockscomb prickleback *Anoplarchus purpurescens* Gill. M.Sc. Thesis, University of British Columbia, Vancouver. BC 159 p.

Polis, G.A. and Hurd, S.D. 1996. Linking marine and terrestrial food webs: allochthonous input from the ocean supports high secondary productivity on small islands and coastal land communities. The American Naturalist. 147(3): 396-423.

Rangeley, R.W. 1994. The effects of seaweed harvesting on fishes: a critique. Environmental Biology of Fishes. 39: 319-323.

Regional District of Nanaimo. 2017. http://www.rdn.bc.ca/cms.asp?wpID=411

Richards, L.J. 1984. Field studies of foraging behaviour of an inter-tidal beetle. Ecological Entomology. 9(2): 189-194.

Rodhouse, P.G. and Roden, C.M. 1987. Carbon budget for a coastal inlet in relation to intensive cultivation of suspension-feeding bivalve molluscs. Marine Ecology Progress Series. 36: 225-236.

Romanuk, T.N. and Levings, C.D. 2003. Associations between arthropods and the supralittoral ecotone: dependence of aquatic and terrestrial taxa on riparian vegetation. Environmental Entomology. 32: 1343-1353.

Romanuk, T.N. and Levings, C.D. 2005. Stable isotope analysis of trophic position and terrestrial vs. marine carbon sources for juvenile Pacific salmonids in near shore marine habitats. Fisheries Management and Ecology. 12(2): 113-121.

Romanuk, T.N. and Levings, C.D. 2006. Relationships between fish and supra-littoral vegetation in near shore marine habitats. Aquatic Conservation: Marine and Freshwater Ecosystems. 16(2): 115-132.

Rossi F., Olabarria, C., Incera, M. and Garrido, J. 2010. The trophic significance of the invasive seaweed *Sargassum muticum* in sandy beaches. Journal of Sea Research 63(1): 52-61.

Schmidt, A.L., Coll, M., Romanuk, T.N. and Lotze, H.K. 2011. Ecosystem structure and services in eelgrass *Zostera marina* and rockweed *Ascophyllum nodosum* habitats. Marine Ecology Progress Series. 437: 51-68.

Shaffer, J.A., Doty, D.C., Buckley, R.M. and West, J.E. 1995. Crustacean community composition and trophic use of the drift vegetation habitat by juvenile splitnose rockfish *Sebastes diploproa*. Marine Ecology Progress Series. 123: 13-21.

Sharp, G.J. and Pringle, J. 1990. Ecological impact of marine plant harvesting in the northwest Atlantic: a review. Hydrobiologia. 204/205: 17-24.

Sisneros, J.A.P., Alderks, W., Leon, K. and Sniffen, B. 2009. Morphometric changes associated with the reproductive cycle and behaviour of the inter-tidal-nesting, male Plainfin midshipman *Porichthys notatus*. Journal of Fish Biology. 74: 18-36.

Smith, B.D., Cabot, E.L. and Foreman, R.E. 1985. Seaweed detritus versus benthic diatoms as important food resources for two dominant subtidal gastropods. Journal of Experimental Marine Biology and Ecology. 92(2-3): 143-156.

Sobocinski, K.L. 2003. The impact of shoreline armouring on supratidal beach fauna of central Puget Sound. MSc. Thesis, University of Washington. 89 p.

Stagnol, D., Renaud, M. and Davoult, D. 2013. Effects of commercial harvesting of inter-tidal macroalgae on ecosystem biodiversity and functioning. Estuarine, Coastal and Shelf Science. 130: 99-110.

Tenore, K.R. 1981. Organic nitrogen and caloric content of detritus. I. Utilization by the deposit-feeding polychaete *Capitella capitata*. Estuarine Coastal Marine Science. 12: 39-47.

Tenore, K.R. 1988. Nitrogen in benthic food chains. pp. 192-206 *In*: Blackburn, T.H. and Sorenson, J. (eds). Nitrogen Cycling in Coastal Marine Environment. John Wiley and Sons Ltd.

Tyron, L. 2012. Ecological concerns regarding proposed beach wrack harvest. Unpublished report to Comox Valley Project Watershed Society, BC. 19 p.

Vandermeulen, H. 2013. Information to support assessment of stock status of commercially harvested species of marine plants in nova scotia: irish moss, rockweed and kelp. DFO Canadian Science Advisory Secretariat Research Document. 2013/042: vi + 50 p.

Zempke-White, W.L., Speed, S.R. and McClary, D.J. 2005. Beach-cast seaweed: a review. New Zealand Fisheries Assessment Report, 2005/44: 47 p.

Chapter 8

Aquaculture in Baynes Sound

Shelley McKeachie

Past Chair/Director, Association for Denman Island Marine Stewards Society

Baynes Sound has been ranked second only to the Fraser River Estuary as the most ecologically important region along the west coast of BC and is *"the most important wetland complex on Vancouver Island. It is internationally recognized as important for migratory water birds as well as providing habitat for at least six salmonid species"* (Jamieson et al. 2001). This critical feeding and overwintering area for migratory birds is designated an Important Bird Area (IBA) (Jamieson et al. 2001, Chapter 6). The Sound is also a significant herring spawning area and nursery for juvenile herring from adjacent Lambert Channel (Chapter 4). Baynes Sound was designated an Ecologically and Biologically Significant Area (EBSA) by Fisheries and Oceans Canada in 2012 (DFO 2013).

In spite of the ecological importance of Baynes Sound, a major push to expand the shellfish industry began in 1998 when the government of British Columbia decided to double, by 2008, the amount of foreshore and offshore crown land allocated to shellfish aquaculture. Since that time, the traditional 'mom & pop' shellfish operations have developed into multi-acres of mechanized, industrial-scale, high density beach, raft, and longline operations. This expansion and industrialization of aquaculture has all been done in the absence of an overall management plan that would take into account the cumulative impacts on the environment, other economic interests, and residents.

INDUSTRY PRACTICES

Baynes Sound produces approximately than 50% of all commercially harvested shellfish in British Columbia (Chapter 4). Ninety percent of the Baynes Sound coastline is under shellfish industry tenure

(Figure 2, Chapter 4). Vast areas of the intertidal are currently under anti-predator netting and are covered by oyster grow-out beds rendering 56% of critical bird habitat inaccessible to both non-migratory and migratory sea ducks and shore birds (Wan and Bendell 2010). Cumulative effects as a result of long-line oyster culture from rafts, barrels and predator netting over clam beds can result in the intertidal and sub-tidal regions being used exclusively for the shellfish industry.

Wan and Bendell (2010), through the use of GIS and historical and current records of bird abundance and distribution on the east side of Vancouver Island within Baynes Sound, showed that regions once identified as areas of high bird use were no longer used, with bird distribution shifting to those regions of the Sound where no aquaculture was present. Of further note, of the various species discussed by Wan and Bendell (2010) between 1976 and 2006, numbers of Bufflehead and Dunlin increased, both Surf and White-winged Scoters numbers remained the same, whereas there were significant declines for the Western Grebe and Pacific Loon. Both the Western Grebe and Pacific Loon occurred in significant numbers prior to the extensive development of the shellfish industry (Wan and Bendell 2010). The Western Grebe is a red listed species – that is, its status is threatened.

Martell (2008) notes that "in addition to habitat loss from urbanization is direct loss of the foreshore due to intertidal shellfish aquaculture. Use of the foreshore exclusively for aquaculture purposes precludes the use of this region for ecologically important roles such as providing key habitat for spawning activities (e.g. the Pacific sand lance [*Ammodytes hexapterus*]), foraging by wildlife and as nurseries."

Other harmful industry practices include: beach alterations such as channeling of fish bearing streams to protect shellfish tenures; driving on the beach resulting in compaction of sediment and destruction of forage fish spawning habitat and salt marshes; creation of berms using beach cobble; use of rebar and Vexar plastic fencing; and the removal of indigenous species like the moon snail and sea stars.

The cumulative impacts of the above-noted aquaculture practices serve to degrade the highly sensitive intertidal habitat, and to decrease marine diversity. These practices destroy habitat for marine birds and other marine life.

Such practices also negatively impact other uses and values such as tourism, recreation, sport and commercial fisheries, and safe anchorage. Abandoned rusted metal structures and rebar placed into the sediment can pose a threat to the casual beachcomber, boaters and swimmers. Improperly secured nets floating in the water column or on the surface of the water, pose entrapment and entanglement hazards to humans and wildlife alike. Aquaculture tenures are now using mechanized equipment such as generators, cranes, power washers on rafts, power augers, and high pressure hoses for geoduck harvest. These create noise, visual and smell pollution. Tenures are often placed adjacent to established residential areas and in regions highly valued for recreation and tourism, providing a classic example of conflicting economic uses.

Industry Debris

Of great concern is the amount of debris generated by the industry. The Association for Denman Island Marine Stewards (ADIMS) and local residents have conducted an annual beach clean-up since 2004, and each year between three to five metric tons of debris are collected, approximately 90% of which are plastics and styrofoam generated by the shellfish industry. The 2018 beach clean-up removed six tonnes of debris, the greatest amount since the cleanup began. Items included oyster blue plastic rope, long rope (plastic), plastic net shell bags, oyster pouches, oyster baskets, rebar and steel (Figure 1).

Figure 1 Example of tonnage of debris collected in the fall 2017 beach clean-up. Approximately 90% of debris is due to the shellfish industry. *(For color image of this figure, see Color Plate Section at the end of the book)*

Plastics and styrofoam are a known threat to marine life and leach toxins into the marine environment. Once the beach clean-up is complete, the debris just continues to wash up year round. What is seen on Denman Island shores is just the tip of the iceberg. There are no enforceable regulations or penalties for this pollution. The industrial debris needs to be stopped at its source.

Geoduck Aquaculture

The green light was given by the Department of Fisheries and Oceans in 2017 for development of both intertidal and deep water geoduck aquaculture. Vast areas of the intertidal and subtidal will be open to geoduck tenures in a region already far beyond capacity (DFO 2017). The combined areas of Baynes Sound and Lambert Channel are the single most important herring spawning area on the BC coast, and possibly the entire Pacific coast. Anti-predator netting, commonly used over both intertidal and subtidal geoduck tenures, poses a threat to herring spawning habitat, in addition to rendering yet more critical bird habitat unavailable to bird and to marine life (Hay 2012). Another potential threat of geoduck aquaculture is the method used for harvesting. Both intertidal and subtidal tenures use air compressors supplying high pressure hoses to liquefy the substrate, changing all sediment geochemical and geophysical properties.

The first intertidal geoduck tenure was installed in 2014 on the west side of Denman Island. The tenure owner and workers used a power auger to 'drill' holes into the sand/gravel substrate and insert lengths of PVC pipe into which the geoduck seed is 'planted' (one geoduck per approx. 18″ length PVC pipe) (Figure 2a). One acre of a typical intertidal geoduck tenure in Puget Sound, Washington, where this type of aquaculture has been conducted for over a decade, uses approximately 43,500 pipes or thirteen km of PVC pipe. (Washington Sierra Club 2018).

Geoduck farming introduces plastics and hence contaminants into the marine environment, results in habitat loss within the highly valued coastal intertidal region, and could potentially facilitate the transfer of disease and pathogens in coastal environments. This could lead to increased marine wildlife mortalities but also, and as noted by Vethaak and Leslie (2016), could increase the risk of exposure to humans thereby impacting human health. Vethaak and Leslie (2016) conclude by noting "…Plastic debris is a notorious marine issue, but we have touched on evidence here indicating it should now be recognized as an emerging *human health issue* as well".

Plastics

Plastics, including PVC pipe, break down over time, and in doing so, leach toxins and particulates into the marine environment (Figure 2b). Vethaak and Leslie (2016) outline three mechanisms by which persistent plastic waste present significant health risks to humans;

1. **Direct toxicity** of the plastic particles themselves e.g., oxidative stress, cell damage, inflammation and impairment of energy allocation functions.

2. **Chemical toxicity** of the plastic debris. This is of direct relevance to current geoduck farming practices as the PVC piping that is used to house the geoduck can contain a number of additives. These can include heat stabilizers, UV stabilizers, and plasticizers, processing aids, impact modifiers, thermal modifiers fillers, flame retardants, biocides and smoke suppressors. PVCs also contain phthalate plasticizers to improve performance. PVC piping is now a shellfish "debris" item on the shores of Baynes Sound where it is mechanically broken down into increasingly smaller pieces. By doing so, the chemical toxicity of the tubing becomes increasingly of concern as the smaller particles can be ingested by marine organisms.

3. **By acting as substratum**, plastic particles provide the vector for pathogenic micro-organisms and parasites (e.g., *Escherichia coli*, *Bacillus cereus* and *Stenotrophomonas maltophila*). As with 2) above, this is of major concern in regard to current geoduck farming practices. The fragmented piping will facilitate the growth of pathogens and parasites enabling the spread of both disease and infections within near shore environments.

(a)   **(b)**

Figure 2a&b   (a) Installation of geoduck piping and
(b) Debris left behind post installation.
http://protectourshorelinenews.blogspot.com/2012/06/pvc-debris-from-corpo-
rate-geoduck.html
(For color image of this figure, see Color Plate Section at the end of the book)

CONCLUSION

The Department of Fisheries and Oceans Canada (DFO) decision to approve geoduck aquaculture on our BC coast will result in the introduction of vast new quantities of toxic PVC plastic into an ocean that is already inundated with plastics. This decision coincided with the Canadian federal government and nations around the world acknowledging the serious threat plastics pose to the health of our oceans and, indeed, the emerging human health issue. In light of this, it makes no sense to proceed with the geoduck aquaculture plan. This decision should be reversed and efforts directed toward developing safe, non-toxic alternatives to plastic aquaculture equipment.

The industrialization and expansion of aquaculture have had significant negative impacts to the beaches and waters in Baynes Sound in the past two decades. Local, provincial and federal governments have policies that appear to deal with these issues but there is little or no enforcement of these potentially protective measures. It is essential that Baynes Sound be designated a Marine Protected Area (MPA), requiring an overall management plan is developed and implemented to ensure that the diverse marine values of Baynes Sound are preserved and protected for the benefit of all users and future generations, rather than appropriated for the benefit of one industry.

REFERENCES

DFO. 2013. Evaluation of proposed ecologically and biologically significant areas in marine waters of British Columbia. DFO Canadian Science Advisory Secretariat Science Advisory Report 2012/075.

DFO. 2017. Integrated Geoduck Management Framework 2017 Pacific Region. http://waves-vagues.dfo-mpo.gc.ca/Library/40596862.pdf

Hay, D. 2012. Aquaculture impacts on herring in Lambert Channel. https://adimsblog.files.wordpress.com/2013/12/aquaculture-i mpacts-on-herring-in-lambert-channel-report.pdf

Jamieson, G.S., Chew, L., Gillespie, G., Robinson, A., Bendell-Young, L., Heath, W., Bravender, B., Tompkins, A., Nishimura, D. and Doucette, P. 2001. Phase 0 review of the environmental impacts of intertidal shellfish aquaculture in Baynes Sound. Canadian Science Advisory Secretariat Research Document 2001/125, ISSN 1480-4883, Canada. [www document]. URL http://www.dfompo.gc.ca/csas/

Martell, A.E. 2008. Trends in bird populations in the comox valley, British Columbia, Canada, from 19762006. BC Birds 18: 2-14

Vethaak, A.D. and Leslie, H.A. 2016. Plastic Debris is a human health issue. Environmental Science and Technology. 50(13): 6825-6826. http://dx.doi.org/10.1021/acs.est.6b02569

Wan, P. and Bendell, L.I. 2011. Use of aerial photography and GIS to estimate the extent of an anthropogenic footprint. Journal of Coastal Conservation. 15: 417-431. DOI: 10.1007/s11852-010-0101-8.

Washington State Sierra Club. 2018. https://www.sierraclub.org/washington/shorelines-marine-ecosystems. Accessed November 3, 2018.

Chapter 9

Legacy and Emerging Pollutants in Marine Mammals' Habitat from British Columbia, Canada: Management perspectives for sensitive marine ecosystems

Juan José Alava[1,2,3]

[1]Institute for the Oceans and Fisheries, University of British Columbia, 2202 Main Mall, Vancouver, British Columbia V6T 1Z4, Canada.
[2]Adjunct Professor, School of Resource and Environmental Management Faculty of Environment, Simon Fraser University, Burnaby, British Columbia, Canada
[3]Fundación Ecuatoriana para el Estudio de Mamíferos Marinos (FEMM), Ecuador

INTRODUCTION

The Canadian west coast is recognized for the wide diversity of marine habitats such as long, deep fjords and channels, protected coastal seas, outer continental shelf areas with submarine canyons and offshore pelagic waters (Ford et al. 2010a). In particular, the marine environment of British Columbia (BC) is home to 31 species of marine mammals of which 25, 5 and 1 are cetaceans, pinnipeds and mustelid, respectively (Ford et al. 2010a; Ford 2014). This region encompasses several areas and critical habitats used as feeding and breeding grounds by marine mammals. To support the biodiversity and populations of marine mammals in BC offshore and inshore waters, it is a paramount to preserve healthy and suitable habitats. However, despite having the longest coastal zone and shoreline in the world, only 1.3% of coastal waters and zone are protected in Canada, in comparison to 30 and 33% of protected coastal

Corresponding Address: Institute for the Oceans and Fisheries, University of British Columbia, 2202 Main Mall, Vancouver, British Columbia V6T 1Z4, Canada. *E-mail*: j.alava@oceans.ubc.ca

zone and ocean environment as Marine Protected Areas (MPAs) in USA and Australia, respectively (CPAWS 2014). This situation leaves the Canadian coastal ecosystems vulnerable to increasing development and industrialization (Sterrit and Uehara 2013).

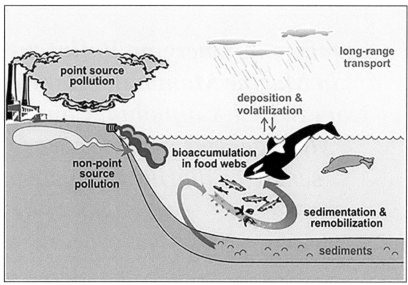

Figure 1 Persistent organic pollutants (POPs, including PCBs and PBDEs), oil spills and hydrocarbons (e.g., petrogenic PAHs), and chemicals of emerging concern (pharmaceuticals and personal care products [PPCPs], and radionuclides) partition among compartments in the marine environment and some of them can readily bioaccumulate and magnify (e.g., POPs) in marine mammalian food webs. Adapted from Alava et al. (2012a).

In addition to the lack of coastal and marine protected areas, the coastal waters of BC are particularly impacted by many anthropogenic activities such as shipping and transportation, pulp mill discharges, and mining and municipal wastewater effluents. This is primarily because the coastal waterways of BC are surrounded by the large urban centres of metro Vancouver harboring a population of ~2.4 million persons (Statistics Canada 2014), and Victoria with a population of 357,327 (Statistics Canada 2014). The marine sediments in this region provide a record of historical and chronic contamination, as they represent a 'sink' for a variety of contaminants, including heavy metals such as lead, mercury and cadmium (Macdonald et al. 1991; Johannessen et al. 2005; Long et al. 2005; Bendell 2010), and persistent organic pollutants (POPs) such as dioxins and furans (Macdonald et al. 1992; Long et al. 2005), polychlorinated biphenyls (PCBs) and the flame retardant polybrominated diphenyl ethers (PBDEs) (Ikonomou et al. 2002;

Rayne et al. 2004; Johannessen et al. 2008; Long et al. 2005), as well as polycyclic aromatic hydrocarbons (PAHs) (Long et al. 2005; Yunker and Macdonald 2003), and industrial detergents (Shang et al. 1999), as illustrated in Figure 1.

While PCBs are not used anymore and declining in the marine environment within the coastal and marine environment of BC, PBDE flame retardants are reaching steady state or declining in biota over time and in the vicinity of BC, as observed in harbour seal pups (Ross et al. 2013; Noël and Ross 2018). However, these POPs are the dominant toxicological concern affecting resident killer whale and continue to biomagnify in regional marine mammalian food webs (Cullon et al. 2005 and 2012; Alava et al. 2012a&b; Alava et al. 2016).

Recent oil pipeline projects in the BC coastal-marine region cannot be ruled out as potential threats as either a direct impact from coastal-based infrastructural operations or maritime traffic of tankers for marine mammals and their habitats. The controversial Enbridge Northern Gateway mega-pipeline project in Northern BC, which was initially approved on 17 June 2014 by the National Energy Board (NEB), emerged as a protracted risk for the permanent and temporal habitats of several species of marine mammals, including the regional population of humpback whales, which is recently recovering from dwindling numbers (Alava and Silberg 2014). Fortunately, this project was rejected and ceased later on by the Canadian Government because of the immediate long-term impacts and high probability of pollution risks (i.e. oil spills) in the sensitive ocean environment along the northern coast of BC. However, the NEB and Government's approval of the Trans-Mountain Kinder Morgan Pipeline project in BC on November 29, 2016 (David Suzuki Foundation 2016; Whittingham 2016) emerged as a new looming threat for marine mammals and their habitats not only due to the potential increase in greenhouse gas emissions, but because of a series of cumulative impacts to the fragile population of harbor porpoises (*Phocoena phocoena*) (Fisheries and Oceans Canada 2009) and the most endangered population of marine mammals, the 74 southern resident killer whales (*Orcinus orca*) (Young and MacDuffee 2018), on BC's south coast habitats (Alava and Calle 2017). The cumulative impact includes increases in tanker traffic by seven fold, acoustic pollution and risks of bitumen oil spills (Alava and Calle 2017). As August 30, 2018, the Federal Court of Appeal halted the government's approval for the construction of the Trans Mountain pipeline and oil tanker expansion as it broke the law and violated Indigenous rights of First Nations and failed to include the negative effects of a sevenfold increase in tanker traffic on the marine environment of killer whales.

Impact assessment and source control programs have received little attention and need to be in place to ensure the protection of marine mammals in coastal BC. This is of important consideration as the Canadian *Species at Risk Act* (SARA) protects species at risk from being killed or harmed (section 32) and protects any part of their Critical Habitat from destruction (section 58). Critical Habitat is defined as the habitat necessary for survival or recovery of a listed wildlife species at risk, as identified in a Recovery Strategy or Action Plan. For instance, one of the major objectives (i.e. Objective 2) of the resident killer whale recovery strategy (Fisheries and Oceans Canada 2008) is to *"ensure that chemical and biological pollutants do no prevent the recovery of resident killer whale populations."* Under SARA, Critical Habitat is legally protected from destruction, and advice from science is needed to justify management decisions designed to protect all of resident killer whale habitat under the *Fisheries Act*.

With the aim to provide evidence and a rationale to preserve marine mammals' habitats and coastal zones in BC, a commitment to assess and predict the impact of pollutants in regional environments to improve risk management tools is required. Importantly, continued monitoring of sensitive coastal marine environments, including Baynes Sound, is required. Within this context, this chapter is a general overview of habitat status and distribution for major marine mammal species inhabiting BC with special emphasis in Baynes Sound and we provide an impact assessment of the anthropogenic chemical pollution affecting these species.

KEY MARINE MAMMAL SPECIES AND HABITATS RELEVANT TO THE VICINITY OF BAYNES SOUND IN BRITISH COLUMBIA

Pacific Harbour Seal (*Phoca vitulina richardsii*)

The harbor seal is likely to be the most abundant and common marine mammal species in BC, where it occurs mainly throughout the inshore and coastal waters of the Strait of Georgia, Johnstone Strait, and west coast of Vancouver Island as well as offshore, remote areas, including Haida Gwaii (Queen Charlotte Islands) and nearby marine environments. The harbor seal population in the BC coast is 105,000 individuals, from which close to 40% is found within the Strait of Georgia (Fisheries and Oceans Canada 2010a). In the Strait of Georgia, harbour seals feed on locally and seasonally abundant prey, including schooling fishes such as Pacific herring (*Clupea pallasi*), Pacific hake (*Merluccius productus*), Pacific sandlance (*Ammodytes hexapterus*), Pacific salmon (*Oncorhynchus* spp.), surfperches (*Embiotocidae*), smelts (eulachon, *Osmeridae*) and walleye pollock (*Theragra chalcogramma*), and bottom fish such as Plainfin midshipman (*Porichthys notatus*), lingcod (*Ophiodon elongatus*), flounder/flatfish (*Pleuronectiformes*), rockfish (*Scorpaenidae*), as well as squid (Olesiuk 1993; Fisheries and Oceans Canada 2010a). Pacific herring and Pacific hake account by 75% of the total annual diet (Olesiuk 1993). These diet items play an important role as biovector delivering contaminants (e.g. POPs) through the harbour seals food web in the Strait of Georgia, as documented recently (Cullon et al. 2005 and 2012). Because of its large population, the harbour seal is considered to be *"Not at Risk"* in western Canada (Olesiuk 1999).

Steller Sea Lion (*Eumetopias jubatus*)

The Steller sea lion (*E. jubatus*) is the largest sea lions of the world and a piscivorous pinniped that inhabits the Pacific coastal waters of Canada, the USA and Asia. There are two populations, with the Eastern and Western stocks being genetically distinct and geographically separated at approximately 145 °W longitude (Bickham et al. 1996). Main diet for the eastern population of sea lions includes Pacific herring, Pacific sandlance, rockfish and Pacific salmon (*Oncorhynchus* spp.). While the eastern stock is considered stable or increasing, the western stock has declined by 80% during the last 30 years across its entire range (National Research Council 2003). In addition to the hypotheses involving nutritional stress, i.e. reduction in the quantity and quality of food (low energy food), and shifts in ocean-climate, which might explain this decline (Rosen and Trites 2000; Trites et al. 2007), contaminants have also been suggested as a possible contributing factor (Barron et al. 2003). The total Steller sea lion population in British Columbia during the breeding season is estimated to be approximately 20,000-28,000 individuals, with an overall growth rate of 3.5% per year, and representing 33% of the eastern stock (Olesiuk 2008). Of these, approximately 3000 individuals migrate into the waters off southern Vancouver Island and into the Strait of Georgia (Olesiuk 2004). Although the British Columbia population has been increasing, Steller sea lions are listed as "Special Concern" under the terms of the Species at Risk Act (SARA) because of human disturbance, risk of oil spills and environmental contaminants (COSEWIC 2003a; Olesiuk 2008).

Resident Killer Whales (*Orcinus orca*)

There are three ecotypes of killer whales (*O. orca*) inhabiting BC marine waters, including fish-eating resident killer whales, transient or marine mammals-eating killer whales and offshore killer whales. Resident killer whales are further distinguished as northern resident killer whales (NRKW) that are often found in the waters off northeast Vancouver

Island, BC, and southern resident killer whales (SRKW) that are often found in the waters off southeast Vancouver Island (Figure 1) (Ford et al. 1998). Resident killer whales are fish-eating marine mammals with a strong preference for Chinook salmon (*Oncorhynchus tshawytscha*), the largest species of Pacific salmon (i.e. ≈50 kg) in the Pacific Northwest, accounting for 72% of the resident killer whales' diet (Ford and Ellis 2006; Ford et al. 2010b&c). Secondary fish species making up the rest of resident killer whales' diet include mainly halibut (*Hippoglossus stenolepis*), sablefish (*Anoplopoma fimbria*), lingcod (*O. elongates*) and several other Pacific salmon species such as sockeye (*O. nerka*), pink (*O. gorbuscha*), Coho (*O. kisutch*) and chum (*O. keta*) salmon (Ford and Ellis 2006; Ford et al. 2010b&c). While the northern resident killer whales have a population of ≈200 individuals, the southern residents is declining with a lower population number of 74 individuals. The population decline, mainly in southern resident, has been associated to several anthropogenic factors such as reduction on quantity and quality of prey (i.e. Pacific wild salmon), human disturbances and underwater noise and pollutants (Fisheries and Oceans Canada 2008). Both the NRKW and SRKW populations have respectively been listed as threatened and endangered under the SARA (Government of Canada 2010a&b). Such a listing requires the protection and conservation of habitat required by resident killer whales.

Harbour Porpoise (*Phocoena phocoena*)

Harbour porpoises are one of the smallest cetaceans and coastal inhabitants residing in shallow waters (< 200 m), and commonly distributed throughout BC waters, including the southern straits, mainland inlets and Queen Charlotte Basin, but showing low densities in deep-waters (e.g., central Strait of Georgia) (Baird 2003; COSEWIC 2003b; Williams and Thomas 2007; Fisheries and Oceans Canada 2009). The average population during summertime was estimated in 9000 individuals (Williams and Thomas 2007), although a lower abundance close to 3000 porpoises were reported in the 1990s based on aerial surveys (Calambokidis et al. 1997).

The major diet items in the diet of harbor porpoises consist of Pacific herring, Pyschrolutidae (likely blackfin sculpin *Malacocottus kincaidi*), Pacific hake, walleye pollock, Pacific sandlance, Pacific sardine (*Sardinops sagax*), northern anchovy (*Engraulis mordax*), and shiner perch (*Cymatogaster aggregata*) (Nichol et al. 2013). The harbor porpoise is listed as a species of "Special Concern" in Canada's Pacific waters and susceptible to anthropogenic activities, including physical disturbances and acoustic pollution, pollutants, bycatch with gillnets and habitat loss (COSEWIC 2003b; Baird 2003; Fisheries and Oceans Canada 2009).

Pacific White-sided Dolphin (*Lagenorhynchus obliquidens*)

The Pacific white-sided dolphin is one of the most widely distributed and abundant small toothed cetaceans in the North Pacific and it is a year round resident of both pelagic and nearshore waters in BC (Heise 1997a; Ford et al. 2010a; Ford 2014). In BC inshore coastal waters, approximately 25,900 dolphins have been found to reside along the coast (Williams and Thomas 2007), where they feed on at least 13 different prey species, including Pacific salmon, Pacific herring, walleye pollock, shrimp, sablefish, smelt and squid (Heise 1997b). In recent years, there have been new important sightings of this species in groups ranging from 50 to 1000 individuals in the Strait of Georgia and Howe Sound (Birdsall 2008; Ford et al. 2010a; Baker 2013). The Pacific white-sided dolphin is considered by COSEWIC to be Not at Risk in Pacific Canadian waters (Ford et al. 2010a).

Humpback Whale (*Megaptera novaeangliae*)

Humpback whales are one of the most charismatic marine mammals in BC. The current North Pacific humpback whale population is about 18,000 adult individuals and bounced back after being decimated on the BC coast due to commercial whaling in the early 1990s to mid-1960s (Nichol et al. 2002; Fisheries and Oceans Canada 2013; Calambokidis et al. 2008). In 2014, the humpback whale was delisted from the endangered status to a species of "Special Concern" (Alava and Silberg 2014) following the advice by COSEWIC (2011). However, the local population using the BC coastal waters as feeding grounds to forage on Pacific herring and krill and as migrating corridors is still small (2,145 whales) relative to the estimated pre-whaling levels ($\approx$4,000 whales) and facing increasing anthropogenic impacts, including vessel strikes, entanglements, acoustic pollution, oil spills and prey reduction (Fisheries and Oceans Canada 2013; Calambokidis et al. 2008; Ford et al. 2009). Moreover, the recent endorsement of the Northern Gateway oil mega-pipeline project by the Canadian government raised concerns for the protection of humpback whale habitat (Alava and Silberg 2014).

Grey Whale (*Eschrichtius robustus*)

The grey whales inhabiting seasonally BC waters belong to the Eastern North Pacific population, which has approximately 20,000 individuals (Rugh et al. 2008; Fisheries and Oceans Canada 2010b). A small population in the low hundreds known as summer residents uses BC waters as feeding grounds (Calambokidis et al. 2002), including Boundary Bay and Haro and Georgia Straits (Fisheries and Oceans Canada 2010b).

In 2010, there were at least two important sightings of grey whales in and around downtown Vancouver (e.g., False Creek. Jericho Beach, English Bay, Squamish Estuary), where these whales were likely foraging on benthic invertebrates such as crab larvae and amphipods filtered from sediments in shallow waters close to shore after the long migration (BC Cetacean Sightings Network 2010). The likelihood of potential oil spills from oil and gas extraction and the associated shipping traffic is a major threat that could affect coastal benthic feeders such as grey whales (Fisheries and Oceans Canada 2010). The grey whale is listed as species of "Special Concern' in Canada (COSEWIC 2004).

Sea otter (*Enhydra lutris*)

Although sea otters are mainly abundant and distributed along the western coast of Vancouver Island with a population of approximately 4,000 individuals and a smaller population remaining adjacent to the central mainland coast (≈700 sea otters) (Nichol et al. 2009), this species is included here because of its extreme susceptibly to catastrophic marine oil spills and chronic exposure to PAHs in BC (Harris et al. 2011a&b). Oil is identified as the major threat to BC sea otters because of its small population size, limited geographical distribution and presence of the population nearby shipping lanes (Sea Otter Recovery Team 2007). The lesson learned from the 1989 *Exxon Valdez* oil spill in Alaska is a clear reminder of the vulnerability of sea otters to acute whole oil exposure, in which 4000 sea otters died (Garrot et al. 2003). In BC, sea otters strongly rely on the consumption of benthic invertebrates (i.e. Goeduck clam, *Panopea abrupta*, California mussel, *Mytilus californianus*, Turban snail, *Tegula funebralis*, red rock crab, *Cancer productus*) found in the interface water-sediment, leading to potentially important contaminant exposure through the intake of large amount of prey (Harris et al. 2011b), as they required to consume up to 25% of their body weight per day (Riedman and Estes 1990). The BC sea otter population is a species of "Special Concern" and they are protected by the Fisheries Act, and the British Columbia Wildlife Act (COSEWIC 2007).

ANTHROPOGENIC POLLUTANTS

Threats to marine mammals and their critical habitats are numerous and include reductions in food supply (note previous presentation on forage fish), biotoxins, ship collisions, bycatch and entanglements, anthropogenic climate change, potential oil spills and chemical, biological and acoustic pollution. Anthropogenic pollutants are known to be one of the major stressors conspiring against the survival of marine mammals in BC, specially affecting the iconic killer whales (Figure 1). For the purpose of this chapter, we focus on the impact of chemical pollution by legacy and emerging POPs, Pharmaceuticals and Personal Care Products (PPCPs), marine plastic debris and microplastics, as well as radionuclides, described as follows.

Persistent Organic Pollutants (POPs)

The Strait of Georgia is considered a regional "sink" acting as a receiving environment and hosting "hot spots" of contamination for POPs, including legacy PCBs and emerging PBDEs (Johannessen et al. 2008; Grant et al. 2010; Pearce and Gobas 2018). While the average concentrations of ΣPCB (pg/g dry weight, dw) in sediments of the southern resident killer whales' critical habitat are still lower relative to those found in the critical habitat found in Puget Sound, Washington State, USA (Figure 2, Pearce and Gobas 2018; Pearce 2018), this abiotic compartment is a continuous source for uptake and trophic transfer of PCBs to the food web of endangered southern resident killer whales (Alava et al. 2012a; Pearce 2018). PCB patterns in these sediments revealed that the dominant PCB congeners in the highest proportion are as follows: PCB 118 >> PCB 153 > PCB 110 > PCB 52 > PCB 101 (Pearce and Gobas 2018; Pearce 2018). As a result, the long-term exposure to and bioaccumulation of POPs in marine mammals inhabiting this region have been documented elsewhere (Ross et al. 2000, 2004, 2013; Cullon et al. 2005, 2012; Alava et al. 2012a&b, 2016). PCB concentrations measured in adult northern and southern resident killer whales range from 9.3-146 mg/kg lipid weight (lw), with these killer whales among the most PCB-contaminated marine mammals in the world (Ross et al. 2000). High concentrations of PBDEs were also observed in male southern residents ($\approx$1.0 mg/kg lw), and male northern residents (0.20 mg/kg lw), as reported by Rayne et al. (2004). The PCB levels measured in the resident killer whales readily exceed thresholds for the onset of adverse health effects determined for other marine mammals that range from 1.3-17 mg/kg lw in blubber or liver (Ross et al. 1996; Hall et al. 2006; Mos et al. 2010). Based on a habitat-based PCB risk assessment and food web bioaccumulation modeling approach, resident killer whales contain concentrations of PCBs and PBDEs that exceed toxic thresholds in all areas and critical habitats inhabited by them within coastal BC (Alava et al. 2012a, 2016), as shown in Figure 3.

In addition, harbour seals from the Strait of Georgia and Haida Gwaii exhibited PCB mean concentrations ± SE of 1.14 ± 0.30 mg/kg lw and 2.50 ± 0.20 mg/kg lw, respectively (Ross et al. 2004). While concentrations were lower compared to that measured in harbour seal pups from Puget Sound, i.e. 18 ± 3.0 mg/kg lw (Ross et al. 2004), PCB levels are decreasing in the Salish Sea (Ross et al. 2013). In fact, PCB concentrations have been declining since 1984 indicating that regulations phased out in North America have reduced PCB inputs into the local coastal food web, while PBDE concentrations increased exponentially during the period 1984-2003, but appear to be declining since then reflecting the regulations put into place in early 2000s (Noël and Ross 2018). However, the biomagnification of PCBs in the harbor seal food web of the Strait of Georgia highlights the protracted risk of PCB-associated impacts on harbor seal health in this region because of the persistence, toxic and bioaccumulative nature of these compounds. There's been estimated to be approximately 76 kg of PCBs in the biota of the Strait of Georgia, of which 1.6 kg is bioaccumulated by harbor seals (Cullon et al. 2012).

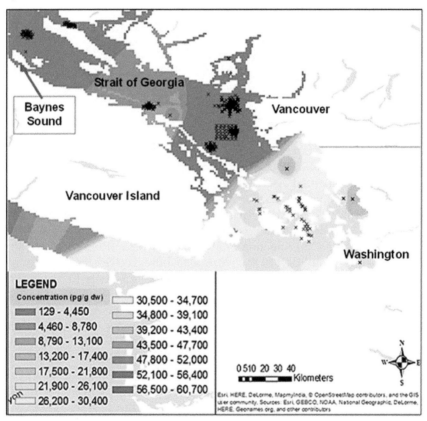

Figure 2 Spatial variation of ΣPCB average concentration (values in pg/g dry weight) and distribution of hot spots in sediments in south BC and Washington estimated using inverse distance weighted (IDW) interpolation method in ArcGIS. Samples sites/locations are indicated with black x. Based on and adapted from Pearce (2018).

(For color image of this figure, see Color Plate Section at the end of the book)

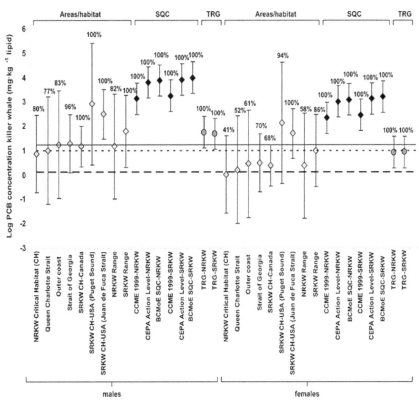

Figure 3 Predicted total PCB concentrations (geometric means in mg/kg lw) and their 95% confidence intervals in males and females from the northern resident killer whales (NRKW) and southern resident killer whales (SRKW) resulting from ΣPCB concentrations in sediment in critical habitats (CH) and areas of BC. The dashed, dotted, and solid lines in ascending order are toxicity threshold concentrations of 1.3, 10, and 17 mg/kg lw, respectively. Label numbers list the proportion of the resident killer whale population exceeding the threshold concentrations of 1.3 mg/kg lw for immunotoxicity and endocrine disruption (Mos et al. 2010). SQC refers to tested sediment concentrations equivalent to sediment quality criteria (i.e. CCME = The Canadian Council of Ministers of the Environment Interim Sediment Quality Guideline ISQG of 21.5 µg/k dry weight (dw), CEPA= the Canadian Environmental Protection Act Action Level Low for disposal at sea of 100 µg/kg dw, BCMoE= British Columbia's Ministry of Environment (BCMoE) sediment quality criterion of 20 µg/kg dw), as described in Alava et al. (2012a), and, TRG stands for the Environment Canada dietary tissue resident guideline (0.05 mg/kg ww) tested in Alava et al. (2012a).
Adapted from Alava et al. (2012a).

Likewise, Steller sea lions are readily exposed to PBDEs and PCBs as are killer whales and harbor seals. Low to moderate concentrations of PCBs have been observed in Steller sea lions from both the declining western stock (Varanasi et al. 1992; Lee et al. 1996; Krahn 1997; Krahn et al. 2001) and the stable eastern stock (Krahn 1997; Krahn et al. 2001). For instance, PBDEs and PCBs were measured in blubber biopsy samples from 22 live-captured Steller sea lions that had just entered the Strait of Georgia (BC, Canada) for their overwintering feeding season. While PBDEs ranged from 0.05 mg/kg lw in adult females to 3.8 mg/kg lw in subadult individuals, PCBs ranged from 0.30 mg/kg lw in adult females to 14.3 mg/kg lw in subadult individuals (Alava et al. 2012b).

Pharmaceuticals and Personal Care Products

Of emerging concerns are also pharmaceuticals and personal care products (PPCPs), which are defined as substances or mixtures of substances for use in personal hygiene, daily cleansing or grooming. These substances are a broad group of chemicals including ingredients in products such as perfumes, soaps and toothpastes, all falling into one of three regulatory categories in Canada: cosmetics, drugs, or natural health products (Health and Environment Canada 2012). Pharmaceuticals include prescription and over the counter medications. In Canada today, there are approximately 11,910 drug products approved for human use and 1356 for veterinary use (Health Canada 2011). Among PPCPs, triclosan represents a risk for marine mammals as triclosan bioaccumulation has been corroborated by a recent study reporting its detection in plasma (9.0 ng/g wet weight, ww) of a captive killer whale (*O. orca*) fed with a diet of Pacific herring harvested from the coast of British Columbia (Bennett et al. 2009). Likewise, a previous study also confirmed for first time the bioaccumulation of triclosan, with plasma concentration ranging 0.025-0.27 ng/g ww, in bottlenose dolphins (*Tursiops truncatus*) from the southeast coast of U.S. (South Carolina and Florida) (Fair et al. 2009).

Recent contaminant data on PPCPs from a systematic monitoring conducted around Greater Victoria, a city located on the southern tip of Vancouver Island BC, Canada, unveiled the presence of several PCPPs with high concentrations within the untreated sewage and PCPP contamination in sediments surrounding the outfalls (Saunders et al. 2016; Krogh et al. 2017). While reference stations from around the region showed very low concentrations of contamination with almost all PPCP concentrations being below detection limits, tissue samples of resident Northern horse mussels (*Modiolus modiolus*) collected adjacent to one of the major sewage outfalls showed high single sample concentrations of the antimicrobial triclosan (317 ng/g dw), the antibiotic ciprofloxacin (176 ng/g dry weight), as well as the antidepressant sertraline (84.0 ng/g dw) (Krogh et al. 2017). Greater Victoria is unique among its North American and European counterparts for its lack of even primary sewage treatment, but the city has started building a tertiary sewage treatment plant with planned operation by 2020 and once the plant is operational, PPCP concentrations within the treated effluent and within the initial dilution zones are expected to be significantly reduced (Krogh et al. 2017). However, based on the lines of evidences aforementioned, the presence of triclosan in marine top predators underlines the need for monitoring in marine mammals and food web bioaccumulation assessments in coastal-marine ecosystems of BC.

Marine Debris: Plastic and Microplastics

In addition to legacy and emerging contaminants, marine life in general is greatly threatened by marine debris, notably plastics and microplastics. Marine debris and plastics can lead to entanglement and ingestion, provide vectors for persistent pollutants, leach toxic chemicals, and provide a vector for invasive species (Moore 2008; GESAMP 2010). It is known that 26 species of cetaceans can ingest plastic bag, fishing line and other plastic around the global ocean (Moore 2008). While further research to assess the impact of plastics in marine mammals of BC has yet to be conducted, a study mapping and projecting the overlap of marine mammals and marine debris in coastal waters of BC estimated that the abundance of floating debris was 36,000 (95% confidence intervals: 23,000-56,600) pieces in the region (Williams et al. 2011). In that study, the most common type of debris was Styrofoam, followed by plastic bottles and plastic bags. The projections also showed that harbor porpoises (*P. phocoena*) and harbor seals (*P. vitulina*), which are found near urban areas off the south coast, were exposed to the highest risk of interacting with marine debris in both urbanized areas off southern Vancouver Island and remote areas including BC's northern mainland fjord. Steller sea lion (*E. jubatus*) were predicted to overlap most strongly with marine debris in Gwaii Haanas National Park Reserve (Williams et al. 2011). Conversely, there is evidence of interactions between Steller sea lions and plastic netting in Baynes Sound, as shown in Figure 4.

Since 1994, the Great Canadian Shoreline Cleanup (GCSC), a joint conservation initiative of Ocean Wise-Vancouver Aquarium and WWF-Canada, has coordinated monitoring and inventories of solid waste and marine debris and cleanups at the local shorelines of BC. Cigarette and filters (butts) and single use plastic items, such as food wrappers, straws, plastic bags and drink bottles, usually dominate (80-90%) the litter found (Konecny et al. 2018). In the southern region of the Strait of Georgia, which includes larger urban areas like Vancouver and Victoria, cigarettes and cigarette filters, which are made of plastic, account for almost 50% of litter recovered (Konecny et al. 2018). However, on the exposed coast of British Columbia, including the north coast of BC (i.e. Haida Gwaii and Prince Rupert), litter composition tends to be different with large plastic bottles and plastic bag being the dominant items. As this coast is exposed to the Pacific Ocean such items are from all over the world and wash up on this shoreline, including some single use plastics, but also large quantities of plastic from fishing, shipping and improperly managed land-based litter (Kate Le Souef, pers. comm. Ocean Wise-Vancouver Aquarium). Natural disasters, such as the Fukushima Earthquake in 2011, have also contributed to shoreline litter in Canada in the form of 'tsunami debris' Kataoka et al. (2017). As an example, 19 cleanups took place in 2013 on exposed west coast, removing 2,295 kg of litter, while 22 cleanups were deployed in 2014, removing 13,860 kg of litter. In 2015, the debris collected from 31 cleanups increased by16,916 kg of litter (Kate Le Souef, pers. comm. Ocean Wise-Vancouver Aquarium). The GCSC recovered a total of 109,557 kg of trash (27,389 ± 66,634 kg/year) in BC between 2013 and 2016 (Konecny et al. 2018).

Figure 4 A male Steller sea lion (*E. jubatus*) interacting with floating marine debris (plastic netting) in inshore waters of Baynes Sound, Vancouver Island, BC. Photo credit: courtesy of Denman Island resident (Association of Denman Island Marine Stewards-ADIMS).

Of particular concern are microplastics, which are defined as particles < 5 mm and can be deliberately manufactured (plastic resin pellets and powder) or generated as break down by-products of larger debris and macroplastic (e.g. clothing, ropes, bags, bottles) (Moore 2008; GESAMP 2010). The final fate of most plastics is derived from land- based sources, including household and industrial water, aquaculture, shipping and tourism.

Desforges et al. (2014) reported that the highest concentrations of microplastics, ranging 4000-5000 particles/m^3 (i.e. detected particles >62 μm), are found in subsurface waters (4.5 m) within the Strait of Georgia, notably in and around Baynes Sound. A high fibre content (% fibre) was observed at nearshore stations and decrease with distance offshore (Desforges et al. 2014). The extent of microplastic pollution impacts on marine biota was also assessed in two species of zooplankton (i.e. the calanoid copepod Neocalanus cristatus and the euphausiid, Euphausia pacifia), resulting in ingestions of 1 particle/every 34 copepods and 1/every 17 euphausiids (Desforges et al. 2015).

In Baynes Sound, microplastic pollution in coastal intertidal sediments with concentrations in the order of 76,500 particles/m^3 has also been documented (Cluzard et al. 2015). Recent data by Kazmiruk et al. (2018) showed that microplastics (i.e. microfibers, microbeads and microfragments) are widely spread and found in all sampling locations within Baynes Sound (i.e. sites 1-6 and 13-16, as shown in Figure 5) and Lambert Channel (sites 7-12, Figure 5). Figure 5 shows that microbeads occurred in the highest number (≈ 25000/kg dry sediment) relative to microfibers and microfragments, which occurred in similar amounts (100-300/kg dry sediment), indicating ubiquitous pollution with these emerging contaminants in this sensitive area (Kazmiruk et al. 2018). These findings provide baseline information that can support risk management and control strategies by authorities and municipalities from cities around the region to reduce and mitigate sources of microplastic pollution in the ocean and impacts in marine biota.

Although research to assess the extent of the impact and health effects of microplastics in marine mammals of the coastal-marine ecosystem of BC, mainly in Sensitive Marine Ecosystems (e.g. Baynes Sound), has yet to be conducted, recent studies show that cetaceans can be exposed and ingest microplastics (Lusher et al. 2015; Fossi et al. 2016). In fact, while ingestion of MPs has been found in the stomach content of a deep diving, oceanic cetacean species (True's beaked whale, *Mesoplodon mirus*) stranded on the coast of Ireland (Lusher et al. 2015), recent research shows that fin whales (*Balaenoptera physalus*) from the Mediterranean Sea are exposed to MPs due to direct ingestion and consumption of contaminated prey (Fossi et al. 2016).

Radionuclides: Impact from the Fukushima Nuclear Accident

Other contaminants that threaten BC coastal waters include radionuclides from the damaged Fukushima plant reactor and debris from the associated tsunami in Japan. The Fukushima nuclear accident on

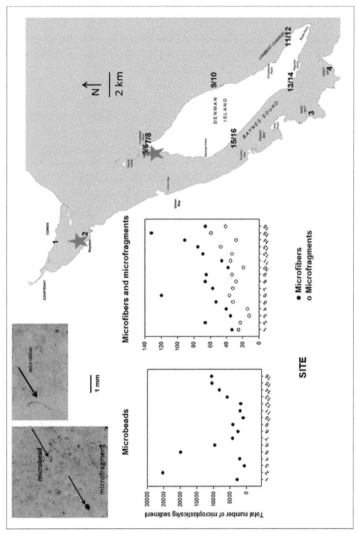

Figure 5 Total microplastic concentrations (particles/kg of dry sediments) observed in 16 sampling sites in Baynes Sounds and Lambert Channel (eastern coast of Vancouver Island, BC, Canada). The inset charts show the number of microplastic particles breakdown as microbeads, microfibers and microfragments per sampling site. The red starts depict the sites with the greatest microbead concentrations in the study area. Adapted from Kazmiruk et al. (2018). Permission to use figures was granted as courtesy of Dr. Leah Bendell (Simon Fraser University, BC, Canada).

11 March 2011 emerged as a global threat for the conservation of the Pacific Ocean, human health, and marine biodiversity (Alava and Gobas 2016). The nuclear emergency situation at Fukushima was initially attributed a severity level 5, equivalent to the Three Mile Island accident in 1979 (i.e. *"an accident with wider consequences"*). Yet, a month later after the earthquake (April 11), the Fukushima nuclear plant reached the severity level 7, equivalent to that of the 1986-Chernobyl nuclear disaster. This accident was defined by the International Atomic Energy Agency (IAEA) as *"a major release of radioactive material with widespread health and environmental effects requiring implementation of planned and extended countermeasures"*. The potential radioactive contamination of seafoods through bioaccumulation of radioisotopes, mainly Cesium 137 (^{137}Cs) with a half-life of 30 years, in marine and coastal food webs are issues of major concern for the public health of coastal communities. Particularly vulnerable were First Nations that rely strongly on the harvest and consumption of traditional seafoods and fish products (e.g. Pacific salmon) (Alava and Gantner 2018). The effects of radioactive contamination are likely to affect other top predators, including fish-eating marine mammals inhabiting offshore and coastal habitats of the region.

Recently, seawater transporting ^{137}Cs associated with the Fukushima accident was detected in waters offshore Vancouver Island (i.e., 1500 kilometres west of BC) at depths of zero to 100 metres starting in 2012. The measured concentrations of ^{137}Cs were very low in 2012 and increased to values in excess of 0.008 becquerels per litre (Bq/L) by 2016 (Smith et al. 2015, 2017).

Although radioactivity levels in fish and seafood products in BC are being considered of no risks for people at present times (Chester et al. 2013; Chen 2013; Chen et al. 2014; Domingo et al. 2016 and 2018; Alava and Gantner 2018), concerns and questions remain about the long-term exposure and bioaccumulation of radioactivity in marine food webs of the North Pacific (Alava and Gobas 2016). A bioaccumulation modeling study to simulate and predict concentrations of ^{137}Cs in the food-web of piscivorous southern resident killer whales from BC in the long term revealed that the ^{137}Cs levels in Pacific herring, sablefish, halibut, Pacific salmon (i.e. pink, chum and Chinook) and killer whales will remain far below the Canada Action Level for food consumption, i.e. 1000 Bq/kg (Alava and Gobas 2016). This threshold is the only benchmark for this radionuclide in Canada to be compared with. In this context, while no radiation risks are expected thus far to the general public, coastal communities and killer whales from the consumption of Pacific salmon and other fish products harvested from marine coastal waters off and marine mammal habitats around BC, environmental radiation is still not well understood (Alava and Gantner 2018).

MANAGEMENT PERSPECTIVES AND IMPLICATIONS FOR SENSITIVE MARINE ECOSYSTEMS

The legacy of contamination and the looming threat of emerging pollutants are continuing to affect and will impact the quality of habitat for marine mammals along coastal BC, particularly in the Strait of Georgia, including Baynes Sound. Management of these contaminants and potential chemical assaults to mitigate their impacts are clearly needed.

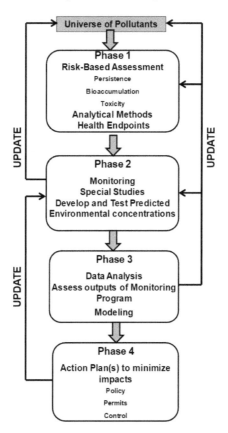

Figure 6 Phases of the adaptive monitoring approach to control and manage pollutants identified and assessed as legacy contaminants and chemicals of emerging concern with the highest risk (phase 1), followed by the development of new monitoring (phase 2) and assessment tools (phase 3), and interpretation of new information to undertake actions and refine the monitoring enterprise (phase 4), as needed. Modified and adapted from Maruya et al. (2014).

The classical notion based on the paradigm that *"The solution to pollution is dilution,"* which is a reactive approach still applied by command and control measures by regulatory agencies to address and mitigate pollution at the end of the pipe must evolve to a proactive approach following the precautionary principle to prevent pollution (i.e. dilution is not the solution to pollution). The recent amendments to the Canadian Environmental Protection Act (CEPA) promotes the adoption of the preventive pathway, but policy and decision makers are in need to internalize a more transparent and comprehensive decision-making processes taking into account the voice of coastal communities and the trade-offs between development and conservation and protection of the marine environment and public health. The adaptive environmental management coupled with a monitoring strategy to prevent and monitor pollution can be an important tool to develop and apply for sensitive marine ecosystems. This adaptive strategy should include four sequential phases that can balance the potential risks identified for chemicals of emerging concern including uncertainty, against escalating actions (Maruya et al. 2014), as seen in Figure 6.

In phase 1, the adaptive approach first identifies these substances as chemicals of emerging concern because of their toxicity, persistence and bioaccumulation potential from the risk assessment based framework, followed by initial monitoring at appropriate spatial and temporal scales using robust analytical methods and evaluates emerging monitoring and assessment technologies (phase 2). Then, it analyzes and interprets initial monitoring data using the most current information and modeling tools (phase 3), and, subsequently, the approach implements control actions commensurate with potential risks in phase 4 (Figure 6). Under this premise, we emphasize the inclusion of an adaptive monitoring and a management approach for legacy and chemicals of emerging concern within the Baynes Sound region. The adaptive monitoring strategy will allow for future review of, and if warranted, modifications to the action strategies based on the feedback resulting from the monitoring to incorporate the latest science available to the natural resources community and decision makers.

Baynes Sound is an important study case and model area, in which proactive actions and prevention of anthropogenic pollution based on adaptive management can be formulated and implemented to manage and preserve its semi-pristine and resilient environment still remaining in the BC marine region.

ACKNOWLEDGEMENTS

The author thanks Dr. Leah Bendell, Laurie Wood and Dr. Patricia Gallagher for organizing and the personal invitation to participate in the workshops on *"Management of sensitive marine ecosystems: Lessons from case studies to identify solutions for Baynes Sound,*" held at Simon Fraser University on 3-4 April, 2014, and to contribute with this chapter.

REFERENCES

Alava, J.J., Ross, P.S., Lachmuth, C.L., Ford, J.K.B., Hickie, B.E. and Gobas, F.A.P.C. 2012a. Habitat-based PCB Environmental Quality Criteria for the Protection of Endangered Killer Whales (*Orcinus orca*). Environmental Science and Technology 46: 12655-12663.

Alava, J.J., Lambourn, D., Olesiuk, P., Lance, M., Jeffries, S., Gobas, F.A.P.C. and Ross, P.S. 2012b. PBDE flame retardants and PCBs in migrating steller sea lions (*Eumetopias jubatus*) in the Strait of Georgia, British Columbia, Canada. Chemosphere 88: 855-864.

Alava, J.J. and Silberg, J.N. 2014. Delisted whales good news for pipeline. Science 345(6194): 278-279.

Alava, J.J. and Gobas, F.A. 2016. Modeling ^{137}Cs bioaccumulation in the salmon–resident killer whale food web of the northeastern pacific following the Fukushima nuclear accdent. Science of the Total Environment 544: 56-67.

Alava, J.J., Ross, P.S. and Gobas, A.P.C. 2016. Food web bioaccumulation model for resident killer whales from the northeastern pacific ocean as a tool for the derivation of PBDE-sediment quality guidelines. Archives of Environmental Contamination and Toxicology 70(1): 155-168 DOI: 10.1007/s00244-015-0215-y

Alava, J.J. and Calle, N. 2017. Pipelines imperil Canada's ecosystem. Science 6321(355): 140-141.

Alava, J.J. and Gantner, N. 2018. Risks and radiation impacts in the BC coastal-marine environment following the Fukushima nuclear accident. pp. 107-116. *In*: Bodtker, K. (ed.). Ocean Watch. Coastal Ocean Research Institute, Ocean Wise Conservation Association-Vancouver Aquarium. Vancouver, British Columbia, Canada.

Baird, R.W. 2003. Update COSEWIC status report on the harbour porpoise *Phocoena phocoena* (Pacific Ocean population) in Canada, in COSEWIC assessment and update status report on the harbour porpoise *Phocoena phocoena* (Pacific Ocean population) in Canada. Committee on the Status of Endangered Wildlife in Canada. Ottawa. 1-22 pp.

Baker, P. 2013. Rare sighting of 1,000 dolphins off the Gulf Islands. Global BC, 1 November 2013. http://globalnews.ca/news/941166/rare-sighting-of-a-1000-dolphins-off-the-gulf-islands/

Barron, M.G., Heintz, R. and Krahn, M.M. 2003. Contaminant exposure and effects in pinnipeds: implications for Steller sea lion declines in Alaska. Science of the Total Environment 311: 111-133.

BC Cetacean Sightings Network. 2010. Grey whale in the city again. Wild Whales, BCCSN, Vancouver Aquarium–Fisheries and Ocean Canada, 1 September 2010. Vancouver, BC. http://wildwhales.org/2010/09/grey-whale-in-the-city-again/

Bendell, L.I. 2010. Cadmium in shellfish: The British Columbia, Canada experience-A mini review. Toxicology Letters 198: 7-12.

Bennett, E.R., Ross, P.S., Huff, D., Alaee, M. and Letcher, R.J. 2009. Chlorinated and brominated organic contaminants and metabolites in the plasma and diet of a captive killer whale (*Orcinus orca*). Marine Pollution Bulletin 58(7): 1078-1083.

Bickham, J.W., Patton, J.C. and Loughlin, T.R. 1996. High variability for control-region sequences in a marine mammal: implications for conservation and biogeography of Steller sea lions (*Eumetopias jubatus*). Journal of Mammalogy 77: 95-108.

Birdsall, C. 2008. Discovering dolphins in the Strait of Georgia. Wild whales, BC Cetacean Sightings Network, Vancouver Aquarium–Fisheries and Ocean Canada, 31 October 2014. Vancouver, BC. http://wildwhales.org/2008/10/discovering-dolphins-in-the-strait-of-georgia/

Calambokidis, J., Osmek, S. and Laake, J.L. 1997. Aerial surveys for marine mammals in Washington and British Columbia inside waters. Final report for contract 52ABNF-6-0092 to the National Marine Mammal Laboratory, NMFS, NOOA, Seattle. 103 pp.

Calambokidis, J., Darling, J.D., Deecke, V.B., Gearin, P., Gosho, M., Megill, W., Tombach, C.M., Goley, P.D., Toropova, C. and Gisborne, B. 2002. Abundance, range and movements of a feeding aggregation of gray whales from California to southeastern Alaska. Journal of Cetacean Research and Management. 4: 267-276.

Calambokidis, J., Falcone, E.A., Quinn, T.J., Burdin, A.M., Clapham, P.J., Ford, J.K.B., Gabriele, C.M., LeDuc, R., Mattila, D., Rojas-Bracho, L., Straley, J.M., Taylor, B.L., Urbán, J.R., Weller, D., Witteveen, B.H., Yamaguchi, M., Bendlin, A., Camacho, D., Flynn, K., Havron, A., Huggins, J., Maloney, N., Barlow, J. and Wade, P.R. 2008. SPLASH: Structure of populations, levels of abundance and status of humpback whales in the north pacific. Final Report for Contract AB133F-03-RP-00078. U.S. Dept of Commerce. 57 pp.

Chester, A., Starosta., K., Andreoiu, C., Ashley, R., Barton, A., Brodovitch, J.C., Brown, M., Domingo, T., Janusson, C., Kucera, H., Myrtle, K., Riddell, D., Scheel, K., Salomon, A. and Voss, P. 2013. Monitoring rainwater and seaweed reveals the presence of ^{131}I in southwest and central British Columbia, Canada following the Fukushima nuclear accident in Japan. Journal of Environmental Radioactivity 124: 205-213.

Chen, J. 2013. Evaluation of radioactivity concentrations from the Fukushima nuclear accident in fish products and associated risk to fish consumers. Radiation Protection Dosimetry 157(1): 1-5. https://doi.org/10.1093/rpd/nct239

Chen, J., Cooke, M.W., Mercier, J.F., Ahier, B., Trudel, M., Workman, G., Wyeth, M. and Brown, R. 2014. Report on radioactivity measurements of fish samples from the west coast of Canada. Radiation Protection Dosimetry 163(2): 261-266. http://dx.doi.org/10.1093/rpd/ncu150

Cluzard, M., Kazmiruk, T.N., Kazmiruk, V.D. and Bendell, L.I. 2015. Intertidal concentrations of microplastics andtheir influence on ammonium cycling as related to the shellfish industry. Archives of Environmental Contamination and Toxicology 69(3): 310-319.

CPAWS. 2104. Dare to be Deep: Charting Canada's Course to 2020. How Canada can meets its 2020 international marine conservation commitment. Canadian Parks and Wilderness Society. Ottawa, ON. 35 p.

COSEWIC. 2003a. COSEWIC assessment and update status report on the Steller sea lion *Eumetopias jubatus* in Canada. Committee on the Status of Endangered Wildlife in Canada. Ottawa, vii + p. 47. http://www.sararegistry.gc.ca/status/status_e.cfm

COSEWIC. 2003b. COSEWIC assessment and update status report on the harbour porpoise *Phocoena phocoena* (Pacific Ocean population) in Canada. Committee on the Status of Endangered Wildlife in Canada. Ottawa. vi + 22 pp.

COSEWIC. 2004. COSEWIC assessment and update status report on the grey whale (Eastern North Pacific population) *Eschrichtius robustus* in Canada. Committee on the Status of Endangered Wildlife in Canada. Ottawa. vii + 31 pp.

COSEWIC. 2007. COSEWIC assessment and update status report on the sea otter *Enhydra lutris* in Canada. Committee on the Status of Endangered Wildlife in Canada. Ottawa, ON. vii + 36 pp.

COSEWIC. 2011. COSEWIC assessment and status report on the humpback whale *Megaptera novaeangliae* in Canada. Committee on the Status of Endangered Wildlife in Canada. Ottawa. x + 32 pp.

Cullon, D.L., Jeffries, S.J. and Ross, P.S. 2005. Persistent organic pollutants in the diet of harbor seals (*Phoca vitulina*) inhabiting Puget Sound, Washington (USA), and the Strait of Georgia, British Columbia (Canada): A food basket approach. Environmental Toxicology and Chemistry 24: 2562-2572.

Cullon, D.L., Yunker, M.B., Christensen, J.R., Macdonald, R.W., Whiticar, M.J., Dangerfield, N. and Ross, P.S. 2012. Biomagnification of polychlorinated biphenyls in a harbor seal (*Phoca vitulina*) food web from the Strait of Georgia, British Columbia, Canada. Environmental Toxicology and Chemistry 31: 2445-2455.

David Suzuki Foundation. 2016. Expanding pipelines now doesn't make environmental or economic sense. November 29, 2016. http://davidsuzuki.org/media/news/2016/11/expanding-pipelines-now-doesnt-make-environmental-or-economic-sense/ Accessed December 2, 2016.

Desforges, J.P.W., Galbraith, M., Dangerfield, N. and Ross, P.S. 2014. Widespread distribution of microplastics in subsurface seawater in the NE Pacific Ocean. Marine Pollution Bulletin 79(1-2): 94-99.

Desforges, J.P.W., Galbraith, M. and Ross, P.S. 2015. Ingestion of microplastics by zooplankton in the Northeast Pacific Ocean. Archives of Environmental Contamination and Toxicology 69(3): 320-330.

Domingo, T., Starosta, K., Chester, A., Williams, J. and Ross, P.S. 2016. Studying levels of Fukushima-derived radioactivity in sockeye salmon collected on the west coast of Vancouver Island. Radiation Physics and Chemistry. https://doi.org/10.1016/j.radphyschem.2016.12.012

Domingo, T., Starosta, K., Chester, A., Williams, J., Lehnert, J.S., Gantner, N. and Alava, J.J. 2018. Fukushima-derived radioactivity measurements in pacific salmon and soil samples collected in British Columbia, Canada. Canadian Journal of Chemistry 96(2): 124-131. https://doi.org/10.1139/cjc-2017-0272

Fair, P.A., Lee, H.B., Adams, J., Darling, C., Pacepavicius, G., Alaee, M., Bossart, G.D., Henry, N. and Muir, D. 2009. Occurrence of triclosan in plasma of wild Atlantic bottlenose dolphins (*Tursiops truncatus*) and in their environment. Environmental Pollution 157(8-9): 2248-2254.

Fisheries and Oceans Canada. 2008. Recovery strategy for the northern and southern killer whales (*Orcinus orca*) in Canada. Fisheries and Oceans Canada, Ottawa: Species at Risk Act Recovery Strategy Series. 81 pp.

Fisheries and Oceans Canada. 2009. Management Plan for the Pacific Harbour Porpoise (*Phocoena phocoena*) in Canada. Species at Risk Act Management Plan Series. Fisheries and Oceans Canada, Ottawa. v + 49 pp.

Fisheries and Oceans Canada. 2010a. Population Assessment Pacific Harbour Seal (*Phoca vitulina richardsi*). DFO Canadian Science Advisory Secretariat Research Document. 2009/011.

Fisheries and Oceans Canada. 2010b. Management Plan for the Eastern Pacific Grey Whale (*Eschrichtius robustus*) in Canada [Final]. Species at Risk Act Management Plan Series. Fisheries and Oceans Canada, Ottawa. v + 60 pp.

Fisheries and Oceans Canada. 2013. Recovery Strategy for the North Pacific Humpback Whale (*Megaptera novaeangliae*) in Canada. Species at Risk Act Recovery Strategy Series. Fisheries and Oceans Canada, Ottawa. x + 67 pp. http://www.sararegistry.gc.ca/document/default_e.cfm?documentID=1344

Ford, J.K.B. and Ellis, G.M. 2006. Selective foraging by fish-eating killer whales *Orcinus orca* in British Columbia. Marine Ecology Progress Series 316: 185-199.

Ford, J.K.B., Rambeau, A.L., Abernethy, R.M., Boogaards, M.D., Nichol, L.M. and Spaven, L.D. 2009. An assessment of the potential for recovery of humpback whales off the pacific coast of Canada. DFO Canadian Science Advisory Secretariat Research Document. 2009/015. iv + 33 p.

Ford, J.K.B., Abernethy, R.M., Phillips, A.V., Calambokidis, J., Ellis, G.M. and Nichol, L.M. 2010a. Distribution and relative abundance of cetaceans in western Canadian waters from ship surveys, 2002-2008. Canadian Technical Report of Fisheries and Aquatic Sciences. 2913: v + 51 p.

Ford, J.K.B, Wright, B.M., Ellis, G.M. and Candy, J.R. 2010b. Chinook salmon predation by resident killer whales: Seasonal and regional selectivity, stock identity of prey, and consumption rates. Pacific Biological Station: Fisheries and Oceans Canada, DFO. Canadian Science Advisory Secretariat Research Document. 2009/101.

Ford, J.K.B., Ellis, G.M., Olesiuk, P.F. and Balcomb, K.C. 2010c. Linking killer whale survival and prey abundance: food limitation in the oceans' apex predator? Biology Letters 6: 139-142.

Ford, J.K.B. 2014. Marine Mammals of British Columbia. Royal BC Museum Handbook Series, Victoria, BC, Canada. 460 pp.

Fossi, M.C., Marsili, L., Baini, M., Giannetti, M., Coppola, D., Guerranti, C., Caliani, I., Minutoli, R., Lauriano, G., Finoia, M.G. and Rubegni, F. 2016. Fin whales and microplastics: the mediterranean sea and the sea of cortez scenarios. Environmental Pollution 209: 68-78.

Garrott, R.A., Eberhardt, L.L. and Burn, D.M. 1993. Mortality of sea otters in Prince William Sound following the Exxon Valdez oil spill. Marine Mammal Science 9: 343-359.

GESAMP. 2010. IMO/FAO/UNESCO-IOC/UNIDO/WMO/IAEA/UN/UNEP Joint Group of Experts on the Scientific Aspects of Marine Environmental Protection. *In*: Bowmer, T. and Kershaw, P.J. 2010 (eds), Proceedings of the GESAMP International Workshop on plastic particles as a vector in transporting persistent, bio-accumulating and toxic substances in the oceans. GESAMP Rep. Stud. No. 82, 68 pp.

Government of Canada. 2010a. Species Profile: Killer Whale Southern Resident Population. Species at Risk Public Registry. http://www.sararegistry.gc.ca/species/speciesDetails_e.cfm?sid=699.

Government of Canada. 2010b. Species Profile: Killer Whale Northern Resident Population. Species at Risk Public Registry. http://www.sararegistry.gc.ca/species/speciesDetails_e.cfm?sid=698.

Hall, A.J., Mcconnell, B.J., Rowles, T.K., Aguilar, A., Borrell, A., Schwacke, L., Reijnders, P.J. and Wells, R.S. 2006. Individual-based model framework to assess population consequences of polychlorinated biphenyl exposure in bottlenose dolphins. Environmental Health Perspectives 114: 60–64.

Harris, K.A., Yunker, M.B., Dangerfield, N. and Ross, P.S. 2011a. Sediment-associated aliphatic and aromatic hydrocarbons in coastal British Columbia, Canada: Concentrations, composition, and associated risks to protected sea otters. Environmental Pollution 159: 2665-2674.

Harris, K.A., Nichol, L.M. and Ross, P.S. 2011b. Hydrocarbon concentrations and patterns in free-ranging sea otters from British Columbia, Canada. Environmental Toxicology and Chemistry 30(10): 2184-2193.

Health and Environment Canada. 2012. Preliminary Assessment, Triclosan, Chemical Abstracts Service Registry Number 3380-34-5. Available online at http://www.ec.gc.ca/ese-ees/default.asp?lang=En&n=6EF68BEC-1. Accessed on November 25, 2013.

Health Canada. 2001. Drug Product Database (DPD). http://webprod.hc-sc.gc.ca/dpdbdpp/index-eng.jsp

Heise, K.A. 1997a. Life history and population parameters of the Pacific white-sided dolphin (*Lagenorhyncus obliquidens*). Report of the International Whaling Commission 47: 817-825.

Heise, K.A. 1997b. Diet and feeding behaviour of pacific white-sided dolphins (*Lagenorhynchus obliquidens*) as revealed through the collection of prey fragments and stomach content analyses. Report of the International Whaling Commission 47: 807-815

Ikonomou, M.G., Rayne, S., Fischer, M., Fernandez, M.P. and Cretney, W. 2002. Occurrence and congener profiles of polybrominated diphenyl ethers (PBDEs) in environmental samples from coastal British Columbia, Canada. Chemosphere 46: 649-663.

Johannessen, S.C., Macdonald, R.W. and Eek, K.M. 2005. Historical trends in mercury sedimentation and mixing in the Strait of Georgia, Canada. Environmental Science and Technology 39: 4361-4368.

Johannessen, S.C., Macdonald, R.W., Wright, C.A., Burd, B., Shaw, D.P. and van Roodselaar, A. 2008. Joined by geochemistry, divided by history: PCBs and PBDEs in Strait of Georgia sediments. Marine Environmental Research. 66: S112-S120.

Kataoka, T., Murray, C.C. and Isobe, A. 2017. Quantification of marine macro-debris abundance around Vancouver Island, Canada, based on archived aerial photographs processed by projective transformation. Marine Pollution Bulletin 132: 44-51.

Kazmiruk, T.N., Kazmiruk, V.D. and Bendell, L.I. 2018. Abundance and distribution of microplastics within surface sediments of a key shellfish growing region of Canada. PLoS ONE 13(5): e0196005. https://doi.org/10.1371/journal.pone.0196005

Konecny, C., Fladmark, V. and De la Puente, S. 2018. Towards cleaner shores: assessing the great canadian shoreline cleanup's most recent data on volunteer engagement and litter removal along the coast of British Columbia, Canada. Marine Pollution Bulletin 135: 411-417.

Krahn, M.M. 1997. Chlorinated hydrocarbon and DDT analyses of blubber from Steller Sea Lions from southeast Alaska, Chapter 5. *In*: Pitcher, K.W. (ed.), Steller sea lion Recovery Investigations in Alaska 1995-1996. Alaska Department of Fish and Game. <http://www.state.ak.usyadfgywildlifeymmysslzp.htm>.

Krahn, M.M., Beckmen, K.B., Pitcher, P.W. and Burek, K.A. 2001. Population survey of organochlorine contaminants in Alaskan Steller sea lions. Final Programmatic Report for the NFWF Funded Project October 2. National Marine Fisheries Service. Seattle, WA.

Krogh, J., Lyons, S. and Lowe C.J. 2017. Pharmaceuticals and Personal Care Products in Municipal Wastewater and the Marine Receiving Environment Near Victoria Canada. Frontier in Marine Sciences. 4: 415. doi: 10.3389/fmars.2017.00415

Lee, J.S., Tanabe, S., Umino, H., Tatsukawa, R., Loughlin, T.R. and Calkins, D.C. 1996. Persistent organochlorines in Steller Sea Lion (*Eumetopias jubatus*) from the Bulk of Alaska and the Bering Sea, 1976–1981. Marine Pollution Bulletin 32: 535–544.

Long, E.R., Dutch, M., Aasen, S., Welch, K. and Hameedi, M.J. 2005. Spatial extent of degraded sediment quality in Puget Sound (Washington State, U.S.A.) based upon measures of the sediment quality triad. Environmental Monitoring and Assessment 111: 173-222

Lusher, A.L., Hernandez-Milian, G., O'Brien, J., Berrow, S., O'Connor, I. and Officer, R. 2015. Microplastic and macroplastic ingestion by a deep diving, oceanic cetacean: The True's beaked whale, *Mesoplodon mirus*. Environmental Pollution 199: 185-191. doi:10.1016/j.envpol.2015.01.023

Macdonald, R.W., Macdonald, D.M., O'Brien, M.C. and Gobeil, C. 1991. Accumulation of heavy metals (Pb, ZN, Cu, Cd) carbon and nitrogen in sediments from Strait of Georgia, British Columbia, Canada. Marine Chemistry 34: 109-135

Macdonald, R.W., Cretney, W.J., Crewe, N. and Paton, D. 1992. A history of octachlorodibenzo-*p*-dioxin, 2,3,7,8-tetrachlorodibenzofuran, and 3,3',4,4'-tetrachlorobiphenyl contamination in Howe Sound, British Columbia. Environmental Science and Technology 26: 1544-1550.

Maruya, K.A., Schlenk, D., Anderson, P.D., Denslow, N.D., Drewes, J.E., Olivieri, A.W., Scott, G.I. and Snyder, S.A. 2014. An adaptive, comprehensive monitoring strategy for chemicals of emerging concern (CECs) in California's aquatic ecosystems. Integrative Environmental Assessment and Management 10: 69-77.

Moore, C.J. 2008. Synthetic polymers in the marine environment: A rapidly increasing, long-term threat. Environmental Research 108: 131-139.

Mos, L., Cameron, M., Jeffries, S.J., Koop, B.F. and Ross, P.S. 2010. Risk-based analysis of PCB toxicity in harbor seals. Integrated Environmental Assessment and Management 6: 631-640.

National Research Council. 2003. Decline of the Steller Sea Lion in Alaska Waters: Untangling Food Webs and Fishing Nets. National Academic Press, Washington, DC.

Nichol, L.M., Gregr, E.J., Flinn, R., Ford, J.K.B., Gurney, R., Michaluk, L. and Peacock, A. 2002. British Columbia commercial whaling catch data 1908 to 1967: A detailed description of the BC historical whaling database" (Canadian Technical Report of Fisheries and Aquatic Sciences. 2371: vi + 77pp., Fisheries and Oceans Canada, Nanaimo, BC.

Nichol, L.M., Boogaards, M.D. and Abernethy, R. 2009. Recent trends in the abundance and distribution of sea otters (*Enhydra lutris*) in British Columbia. Canadian Science Advisory Secretariat Research Document 2009/016. Fisheries and Oceans Canada, Ottawa, ON. 31pp.

Nichol, L.M., Hall, A.M., Ellis, G.M., Stredulinsky, E., Boogaards, M. and Ford, J.K., 2013. Dietary overlap and niche partitioning of sympatric harbour porpoises and Dall's porpoises in the Salish Sea. Progress in Oceanography 115: 202-210.

Noël, M. and Ross, P.S., 2018. Persistent organic pollutants in marine mammals in British Columbia. *In*: Bodtker, K. (ed.). Ocean Watch. Coastal Ocean Research Institute, Ocean Wise Conservation Association-Vancouver Aquarium. Vancouver, British Columbia, Canada.

Olesiuk, P.F. 1993. Annual prey consumption by harbor seals (*Phoca vitulina*) in the Strait of Georgia, British Columbia. Fisheries Bulletin 91: 491–515.

Olesiuk, P.F. 1999. An assessment of the status of harbour seals (*Phoca vitulina*) in British Columbia. Canadian Stock Assessment Secretariat Research Document 99/33. Ottawa, ON, 71 pp.

Olesiuk, P.F. 2004. Population biology and status of Steller and Californian Sea Lions (*Eumetopias jubatus* and *Zalophus californianus*) in Canadian waters. p. 91. *In*: Abstracts of the Sea Lions of the world: Conservation and Research in the 21st Century-22nd Wakefield Fisheries Symposium. 30 September–3 October, 2004. Anchorage, AK, USA,

Olesiuk, P.F. 2008. Population Assessment: Steller Sea Lion (*Eumetopias jubatus*). DFO Canadian Science Advisory Secretariat Science Advisory Report 2008/047: p. 11.

Pearce, R. 2018. Bioaccumulation of PCBs in Southern Resident Killer Whales in the Salish Sea. MRM Project. Report No. 706. School of Resource and Environmental Management, Faculty of Environment, Simon Fraser University. 115 pp.

Rayne, S., Ikonomou, M.G., Ross, P.S., Ellis, G.M. and Barrett-Lennard, L.G. 2004. PBDEs, PBBs, and PCNs in three communities of free-ranging killer whales

(*Orcinus orca*) from the northeastern pacific ocean. Environmental Science and Technology 38: 4293-4299.

Riedman, M.L. and Estes, J.A. 1990. The sea otter (*Enhydra lutris*): behavior, ecology, and natural history. Biological Report 90(14). U.S. Department of the Interior Fish and Wildlife Service, Washington, DC.

Rosen, D.A.S. and Trites, A.W. 2000. Pollock and the decline of steller sea lions: testing the junk-food hypothesis. Canadian Journal of Zoology 78: 1243-1258.

Ross, P.S., De Swart, R.L., Timmerman, H.H., Reijnders, P.J.H., Vos, J.G., Van Loveren, H. and Osterhaus, A.D.M.E. 1996. Suppression of natural killer cell activity in harbor seals (*Phoca vitulina*) fed Baltic Sea herring. Aquatic Toxicology 34: 71-84.

Ross, P.S., Ellis, G.M., Ikonomou, M.G., Barrett-Lennard, L.G. and Addison, R.F. 2000. High PCB concentrations in free-ranging Pacific killer whales, *Orcinus orca*: Effects of age, sex and dietary preference. Marine Pollution Bulletin 40: 504-515.

Ross, P.S., Jeffries, S.J., Yunker, M.B., Addison, R.F., Ikonomou, M.G. and Calambokidis, J. 2004. Harbour seals (*Phoca vitulina*) in British Columbia, Canada, and Washington, USA, reveal a combination of local and global polychlorinated biphenyl, dioxin, and furan signals. Environmental Toxicology and Chemistry 23: 157-165.

Ross, P.S., Noel, M., Lambourn, D., Dangerfield, N., Calambokidis, J. and Jeffries, S. 2013. Declining concentrations of persistent PCBs, PBDEs, PCDEs, and PCNs in harbor seals (*Phoca vitulina*) from the Salish Sea. Progress in Oceanography 115: 160-170.

Rugh, D.J., Breiwick, J., Muto, M.M., Hobbs, R.C., Sheldon, K.W., D'Vincent, C., Laursen, I.M., Rief, S.L., Maher, S.L. and Nilson, S.D. 2008. Report of the 2006-2007 census of the eastern north pacific stock of gray whales. AFSC Processed Report. 157pp.

Saunders, L.J., Mazumder, A. and Lowe, C.J. 2016. Pharmaceutical concentrations in screened municipal wastewaters in Victoria, British Columbia: a comparison with prescription rates and predicted concentrations. Environmental Toxicology and Chemistry 35: 919-929. doi: 10.1002/etc.3241

Sea Otter Recovery Team. 2007. Recovery Strategy for the Sea Otter (*Enhydra lutris*) in Canada. Species at Risk Act Recovery Strategy Series. Fisheries and Oceans Canada, Vancouver, BC.

Shang, D.Y., Macdonald, R.W. and Ikonomou, M.G. 1999. Persistence of nonylphenolethoxylate surfactants and their primary degradation products in sediments from near amunicipal outfall in the Strait of Georgia, British Columbia, Canada. Environmental Science and Technology 33: 1366-1372.

Smith, J.N., Browns, R., Williams, J.W., Robert, M., Nelson, R. and Moran, S.B. 2015. Arrival of the Fukushima radioactivity plume in North American continental waters. Proceedings of the National Academy of Sciences 112(5): 1310-1315.

Smith, J.N., Rossi, V., Buesseler, K.O., Cullen, J.T., Cornett, J., Nelson, R., Macdonald, A.M., Robert M. and Kellogg, J. 2017. Recent Transport History of Fukushima Radioactivity in the Northeast Pacific Ocean. Environmental Science and Technology 51(18): 10494-10502.

Sterrit, A. and Uehara, M. 2013. Canada must do more to protect coastal waters from increasing industrialization. The Vancouver Sun, 24 March 2013. http://www.vancouversun.com/technology/Canada+must+more+protect+coastal+waters+from+increasing+industrialization/8145802/story.html

Statistics Canada. 2014. Estimates of population by census metropolitan area, sex and age group for July 1, based on the Standard Geographical Classification (SGC) 2011, annual (persons), CANSIM (database) Table 051-0056. Accessed: 2014-06-27 http://www5.statcan.gc.ca/cansim/a26?lang=en g&retrLang=eng&id=0510056&paSer=&pattern=&stByVal=1&p1=1&p2=37&t abMode=dataTable&csid=

Trites, A.W., Miller, A.J., Maschner, H.D.G., Alexander, M.A., Bograd, S.J., Calder, J.A., Capotondi, A., Coyle, K.O., Di Lorenzo, E., Finney, B.P., Gregr, E.J., Grosch, C.E., Hare, S.R., Hunt Jr. G.L., Jahncke, J., Kachel, N.B., Kim, H., Ladd, C., Mantua, N.J., Marzban, C., Maslowski, W., Mendelssohn, R., Neilson, D.J., Okkonen, S.R., Overland, J.E., Reedy-Maschner, K.L., Royer, T.C., Schwing, F.B., Wang, J.X.L. and Winship, A.J. 2007. Bottom-up forcing and the decline of steller sea lions (*Eumetopias jubatus*) in Alaska: assessing the ocean climate hypothesis. Fisheries Oceanography 16: 46-67.

Varanasi, U., Stein, J.E., Reichert, W.L., Tilbury, K.L., Krahn, M.M. and Chan, S.L. 1992. Chlorinated and aromatic hydrocarbons in bottom sediments, fish and marine mammals in US coastal waters: laboratory and field studies of metabolism and accumulation. pp. 83-115. *In*: Walker, C.H. and Livingstone, D.R. (eds), Persistent Pollutants in Marine Ecosystems. Pergamon Press, Oxford.

Williams, R. and Thomas, L. 2007. Distribution and abundance of marine mammals in the coastal waters of British Columbia, Canada. Journal of Cetacean Research and Management 9(1): 15-28.

Williams, R., Ashe, E. and O'Hara, P.D. 2011. Marine mammals and debris in coastal waters of British Columbia, Canada. Marine Pollution Bulletin, 62(6): 1303-1316.

Whittingham, E. 2016. Pipeline approvals highlight need for a complete pan-Canadian plan. Pembina Institute reacts to approval of the Kinder Morgan Trans Mountain Expansion and Enbridge Line 3 replacement projects. Pembina Institute. November 29, 2016. https://www.pembina.org/media-release/pipeline-approvals-highlight-need-for-a-complete-pan-canadian-plan Accessed December 2, 2016.

Young, J. and MacDuffee, M.2018. Salish Sea orcas need immediate actions to survive. David Suzuki Foundation. June 26, 2018. https://davidsuzuki.org/expert-article/salish-sea-orcas-need-immediate-actions-to-survive/

Yunker, M.B. and Macdonald, R.W. 2003. Alkane and PAH depositional history, sources and fluxes in sediments from the Fraser River Basin and Strait of Georgia, Canada. Organic Geochemistry 34: 1429-1454.

Chapter 10

The Economic Value of New Technologies: Integrated multi-trophic aquaculture in BC

Duncan Knowler

Associate Professor, School of Resource and Environmental Management, Simon Fraser University, Burnaby BC

INTRODUCTION

As conflicts in coastal areas intensify, there is an increasing demand for new ways of doing things that reduce environmental and social impacts while maintaining the economic benefits from resource use. A case in point: British Columbia's coastline is subject to contested use involving aquaculture, chiefly salmon farming but also shellfish aquaculture (e.g. Baynes Sound). While resolving the problems associated with these contested uses is a long and complex process, one potentially fruitful avenue is the exploration of alternative production technologies that can lessen, if not eliminate, the impacts of concern. Precedents in BC are many, including the displacement of clear cut logging in the forest sector with more sustainable selective cutting methods, the installation of scrubbers on coal-fired power plants that use high-sulfur coal, or the development of advanced reclamation techniques in the open pit mining industry. In this chapter, I consider one such improved technology, integrated multi-trophic aquaculture or IMTA, that holds at least some promise of increasing the sustainability of coastal aquaculture in suitable locations in BC. But to understand its potential, it is helpful to place IMTA in the wider context of technological options for the salmon farming industry, and this chiefly involves consideration of closed containment aquaculture.

While technical issues are obviously critical to the development and assessment of improved production technologies, the economic aspects can be just as important, since without economic viability these new techniques either require extensive government support (perhaps justifiable or perhaps not) or will not be adopted by industry. But if governments are to decide whether to support the adoption of new technology they need to be sure that the gains from doing so outweigh the costs. In other words, is society better off with or without this new technology? As a result, we might ask ourselves a more fundamental question: What is the value of a new technology? To economists, the answer to this question has three components: (i) increased benefits to consumers from consuming the products of the new technology; (ii) private financial benefits for producers using the technology; and (iii) benefits (or decreased costs) accruing to society at large when the new technology is in use (e.g. environmental benefits). We have been examining all three components arising from the potential adoption of improved salmon-based aquaculture methods in Canada with funding from the Canadian Integrated Multi Trophic Aquaculture network (CIMTAN). In this chapter some of the key issues and results arising from these investigations are outlined, beginning with the wider context of competing technologies for sustainable salmon aquaculture production in BC.

IMTA VERSUS CLOSED CONTAINMENT AQUACULTURE

Various options exist for improving salmon aquaculture in BC but by far the most discussed is closed containment aquaculture (CCA), primarily land-based (versus sea-based). A second option has arisen more recently, entitled integrated multi-trophic aquaculture (IMTA) and has garnered less attention thus far. Since IMTA is the topic of our own research, we will consider it first. IMTA integrates fed aquaculture (e.g. finfish) with extractive aquaculture (e.g. shellfish and seaweeds) for a more balanced approach to aquaculture that "mimics" some of the characteristics of a natural ecosystem within a more highly managed agro-ecosystem. In particular, IMTA addresses the issue of nutrient loading at monoculture aquaculture sites (Chopin et al. 2008). Implementing IMTA at finfish aquaculture operations involves placing extractive farmed species, such as shellfish and aquatic plants, in close proximity to finfish cages. As a result, the extractive species can consume portions of the organic and inorganic nutrient wastes from the fish component, as shown below (Figure 1).

Figure 1   Integrated multi-trophic aquaculture (IMTA).
(For color image of this figure, see Color Plate Section at the end of the book)

Increasingly, researchers are arguing for an additional benthic component consisting of lobsters, sea urchins or sea cucumbers that is positioned below the net pens to intercept heavier solid wastes (Cubillo et al. 2016). While our focus is on finfish as the fed species and shellfish and seaweeds as the organic and inorganic extractive aquaculture species, respectively, other species that can be used in IMTA include bivalves, algae and benthic invertebrate species. IMTA can be practiced in aquaculture systems on land or in the sea (Chopin and Sawhney 2009).

To date, only one site in BC has produced finfish and extractive products using IMTA (Kyuquot Seafoods at Kyuquot Sound, Vancouver Island), and the finfish species under culture has been sablefish rather than salmon. However, on the East Coast the piloting of IMTA is more advanced with Cooke Aquaculture having actively experimented at a number of farm sites and having announced in January 2011 that it was selling its IMTA-produced salmon to retail grocer Loblaws in eastern Canada. (see: https://thefishsite.com/articles/store-chain-moves-to-sustainable-salmon - accessed: 05/09/18).

Adopting IMTA on a large scale in BC would involve reconfiguration of the farming system at most sites. In contrast to the circular net-pen configuration used in salmon farming on the east coast of Canada, the standard configuration of 12 net pens laid out in parallel in BC would not be conducive to IMTA production and, instead, a single line of cages would work much better (Figure 2). This allows better alignment of the three components of the IMTA system to optimize the biomitigation service performed by the extractive species (e.g. shellfish and seaweeds).

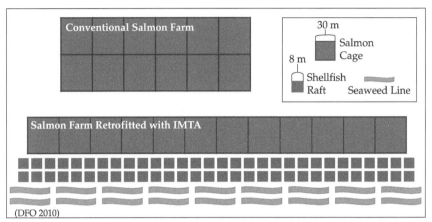

(DFO 2010)

Figure 2 How would an IMTA farm be configured in British Columbia (versus New Brunswick) (Kitchen 2011)?
(For color image of this figure, see Color Plate Section at the end of the book)

As noted above, IMTA is not the only alternative production technology available to improve the sustainability of BC's salmon farming industry. The most highly promoted alternative system for salmon aquaculture in BC is (land-based) closed containment. CCA separates salmon farming from the natural marine environment by using closed salt water tanks on land or in the sea to raise salmon, requiring only minimal or no interaction with the marine environment. CCA can eliminate the nutrient loading, sea lice outbreaks, fish escapes and disease transfer to wild salmon stock commonly associated with open ocean salmon farms (MMK Consulting Inc. 2007; Living Oceans Society 2011). Typical configurations for sea-based and land-based CCA are shown below (Figure 3).

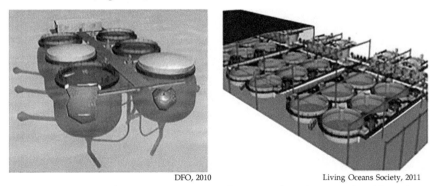

DFO, 2010   Living Oceans Society, 2011

Figure 3 Closed containment aquaculture: sea-based versus land-based.

Support for CCA from several key stakeholder groups has been a dominant feature of the debate thus far, but relatively little research has been undertaken on the economic and market implications of the technology, as with IMTA. Nonetheless, several pilot and pre-commercial sites have produced finfish using CCA in BC: the Namgis Project on northern Vancouver Island has been using a pilot re-circulating aquaculture system since 2014 and initially marketed its limited production to Safeway Sobey's stores in BC and Alberta; AgriMarine Technologies is using a flow-through semi-closed containment system; and Golden Eagle Aquaculture (formerly Swift Aquaculture) has a pilot scale land-based system in operation at Agassiz, BC (For Namgis, see: http://tidescanada.org/wp-content/uploads/2016/10/Namgis_RAS_Salmon_Project_Report_9.pdf; for AgriMarine, see: http://bcsalmonfarmers.ca/b-c-company-leads-aquaculture-industry-with-tech-advancements/; and for Swift, see: https://www.aquaculturenorthamerica.com/profiles/new-management-revitalizes-bc-land-based-coho-farm-1252 (accessed: 05/09/18). While some of these operations have reported promising economic returns, this issue has been hotly debated with competing views reported. For example, see reports by Wright and Arianpoo (2010) and DFO (2013). The Namgis project reports potential improvements in economic returns with increases in operational scale and experience but was experiencing negative gross margins in 2015 (see website above)

WHAT IS THE VALUE OF IMTA AS A NEW TECHNOLOGY?

As indicated earlier, the chapter is concerned with the value of a new technology, which in this case is IMTA in BC. The three areas of potential benefit – for consumers, producers and the public at large – are addressed in the sections below.

(i) What are the benefits of IMTA (versus CCA or wild salmon) for consumers?

Most (about 80 to 85%) of the farmed Atlantic salmon produced in BC is exported, and most of that goes to the US West Coast. Our survey-based research involved assessing preferences for Atlantic salmon produced via IMTA versus CCA in three key "consuming" markets on the US West Coast with a sample size of close to 2000 respondents. After a description of each technology using both balanced and more favorable wordings, responses indicated a preference for IMTA (30% somewhat and 14.2% much more) rather than CCA (13.3% somewhat and 3% much more). Most respondents who preferred IMTA believed it was "more natural" (70% of respondents), "more environmentally friendly"

(60%), "more sustainable" (49%), and "more effective in addressing conventional salmon aquaculture issues" (45%) than CCA. The figure below provides the details.

Such differences in preferences for one technology over the other in the key consuming region for Atlantic salmon from BC can be presented in terms of whether and by how much a price premium could be charged for IMTA or CCA salmon. For example, Barrington et al. (2010) found that focus group participants were willing to pay a 10% premium for labelled IMTA seafood products. An attitudinal study in New York revealed that 38% of the respondents were willing to pay 10% more for IMTA mussels compared to conventionally produced mussels (Shuve et al. 2009), while Kitchen (2011) found that oyster consumers in San Francisco were willing to pay premiums of 24% to 36% for IMTA oysters versus conventionally produced oysters. In contrast, a feasibility study of CCA options for the BC salmon aquaculture industry used "subject matter experts" to determine that salmon produced by a type of closed containment system could generate a premium of CAD \$0.33/lb compared to conventionally produced salmon (DFO 2010).

Using a discrete choice experiment (DCE), which is an attribute-based stated preference valuation method, we estimated the price premium that consumers were willing to pay for farmed salmon using IMTA and CCA in three key US West Coast markets (Seattle, Portland and San Francisco). Consumers were willing to pay average price premiums of 9.7% and 3.9% for IMTA and CCA salmon, respectively, in comparison to conventionally priced farmed salmon. Percentage premiums are based on the reference price of USD \$10.99/lb for conventionally produced Atlantic salmon. It seems clear that in the principal consuming region for farmed salmon produced in BC there is a distinct preference for IMTA over closed containment as a production technology. Our follow up discussions with respondents in our surveys suggested that they viewed IMTA as more ecologically friendly and, compared to consumption of wild caught salmon, having less likelihood of depleting natural salmon populations. Further details of this research can be found in Yip (2012) and Yip et al. (2017). The former is accessible online at http://rem-main.rem.sfu.ca/theses/YipWinnie_2012_MRM530pdf.

We carried out further investigations of consumer willingness-to-pay but focusing on IMTA extractive species. Taking oysters as an example, we used a payment card stated preference valuation approach to determine how oyster consumers valued oysters produced at IMTA sites compared to conventionally produced oysters (Kitchen 2011). See Kitchen (2011) and Kitchen and Knowler (2013 for full details of this study; the former can be found online at http://rem-main.rem.sfu.ca/theses/KitchenPatrick_2011_MRM516.pdf. The results of the study revealed that a small percentage of respondents (6.7%) were only willing to pay less than the conventional price; roughly a quarter of respondents of the sample (23.9%) indicated they would pay the conventional price; while the majority of respondents (69.4%) stated they would pay a premium of at least 10%. Respondents considered the IMTA oyster to be a high-value product and were willing to pay an average of 24% more for it. The more environmentally aware and concerned an individual, the more likely they were to pay a premium for IMTA oysters. Similarly, the more positive their perception of aquaculture, the more likely they were to pay a premium.

(ii) Is IMTA attractive to producers?

An earlier study by Ridler et al. (2007) examined the economic potential for an IMTA system involving salmon-mussels-kelp in New Brunswick. They found that over 10 years with a 5% discount rate the net present value (NPV) for IMTA was 20% higher than that of a comparable salmon monoculture system. Whitmarsh et al. (2006) found similar results for an integrated salmon-mussel system in the UK. For adoption of IMTA in BC by conventional salmon farming operations a substantive reconfiguration of the net pen layout would normally be required, as shown in Figure 2, and this is likely to increase upfront investment costs, perhaps dampening economic returns.

As a comparison, several studies have examined the potential profitability of conventional net-pen operations and CCA in BC and these have shown mixed results. For example, DFO (2010) found that conventional net pens generated an internal rate of return (IRR) of 40.6% versus 3.4% for a recirculating closed containment aquaculture system (RAS); however, Wright and Arianpoo (2010) estimated an annual gross margin for CCA of $5 million (44%) for a 1000 MT operation and somewhat higher, assuming a 25% price premium. A complicating factor is that several cost components were not included in both the Ridler (IMTA) and Wright and Arianpoo (CCA) studies, so more analysis has been needed.

We recently investigated the economics of IMTA in New Brunswick (not BC) but this work should help in clarifying the financial attractiveness of IMTA from a full cost accounting perspective (Carras 2017). Results and full details for this study are reported in Carras (2017 and Carras et al. (submitted). The former is available online at http://rem-main.rem.sfu.ca/theses/CarrasMark_2017MRM671pdf. Using a discounted cash-flow analysis (DCF), we examined the financial returns from investing in: (i) a conventional monoculture with Atlantic salmon; (ii) an IMTA operation with Atlantic salmon, blue mussel (*Mytilus edulis*), and kelp (*Saccharina latissima*); and (iii) an IMTA operation with Atlantic salmon, blue mussel, kelp plus green sea urchin (*Strongylocentrotus droebachiensis*), the latter serving as a benthic component positioned beneath the net pens. We found that an IMTA operation comprising three species was more profitable than both Atlantic salmon monoculture and the four-species IMTA, and that the four-species IMTA has a lower net present value (NPV) than salmon monoculture unless there is a price premium for IMTA salmon and mussels. When a 10% price premium on IMTA salmon and mussels was included there was a substantially higher NPV for three-species and four-species IMTA compared to salmon monoculture. Given that the research reported in the previous section suggested a 9.7% price premium would be viable for IMTA salmon exports from BC, it seems reasonable to view the NPV calculations with a 10% price premium as having more credibility.

While the economic study results described above apply to the east coast of Canada and not the BC coast they are instructive. IMTA seems to provide potential opportunities for aquaculture producers willing to adopt the technology. However, ongoing uncertainty related to IMTA's financial and environmental performance, as well as IMTA's increased operational complexity, may be barriers to IMTA adoption in eastern Canada. Additionally, the relatively minor contribution to overall revenues from the extractive species (mussels and kelp), may negatively influence IMTA's potential adoption in Atlantic Canada's salmon-dominated aquaculture industry. On the BC coast, there seems to be even less enthusiasm for IMTA amongst mainstream Atlantic salmon producers, and this seems to be partially shared by First Nations, environmental advocates and other critics of the BC-based salmon farming industry.

(iii) How does the BC general public value the environmental benefits of IMTA?

Residents of BC who consume BC farmed salmon may care about the benefits from a cleaner environment and may be willing to pay more for IMTA or CCA produced salmon, as was evident in the US consuming region. However, this magnitude will be small, even negligible, since the share of farmed salmon production consumed within the province is small, at less than 10%. More importantly, residents of BC generally care about the environment and the coastal environment in particular, whether they are consumers of farmed salmon or not, and this value may be ignored if we consider only purchasers of farmed salmon. For example, the general public may benefit from an improvement in the marine environment arising from the adoption of more sustainable aquaculture technologies. This general societal benefit from a shift to more sustainable technologies needs to be captured separately from the values expressed via purchasing behavior for more sustainably produced salmon.

To explore these preferences and to assess how much BC residents would be willing to pay to support the adoption of more sustainable aquaculture technologies, we carried out a survey of BC residents mostly, but not all, in the coastal region. The study used a discrete choice experiment (DCE) administered via an online survey of 1321 residents of BC to address several research questions involving the willingness-to-pay for improvements in marine environmental quality and the willingness to support adoption of more sustainable technologies as a means of obtaining these improvements in environmental quality. In an earlier study, Martínez-Espiñeira et al. (2016) used contingent valuation, another stated preference valuation methodology, to estimate the potential benefits of IMTA in Canada from biomitigation of salmon farm waste. Focusing on Canadians who do not eat salmon, this study found that in aggregate these individuals would be willing-to-pay between CDN $43 and 65 million per year for the environmental improvements brought about by IMTA. Our research extended this work but used different valuation methods and focused exclusively on BC as the "producing" region versus the "consuming" region in the western USA discussed earlier. This research is reported in Irwin (2014), available online at http://rem-main.rem.sfu.ca/these/IrwinKimberly_2014_MRM593.ppdf.

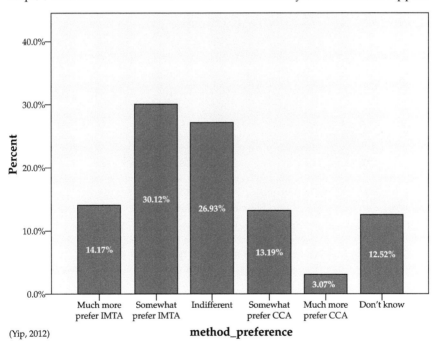

(Yip, 2012)

method_preference

Figure 4   Consumer preferences for IMTA versus CCA in the consuming region (West Coast, USA).
(If either IMTA or CCA was to be adopted for salmon farming, how strong is your preference for one method over the other?)

Attitudes in BC towards IMTA versus CCA indicate a more positive response to CCA than IMTA, when respondents were asked to choose one technology over the other. Approximately one-third (34.7%) of respondents indicated that they somewhat or strongly preferred CCA, while only 25.5% preferred IMTA (see Figure 5). The question asked of BC respondents was identical to the question asked in the USA consuming region (see Figure 4), where the preference clearly was for IMTA over CCA. In contrast, BC respondents just as clearly showed a preference for CCA over IMTA. Respondents preferred CCA because they perceive that it separates farmed salmon from the marine environment (65.1%) and believe it is more effective at addressing salmon farming's environmental issues (60.1%).

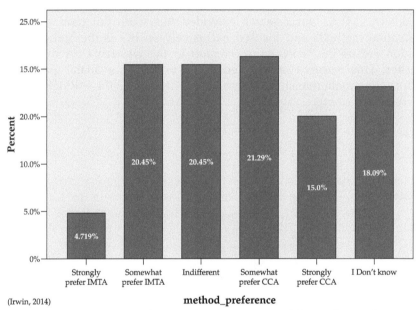

(Irwin, 2014)   **method_preference**

Figure 5 Consumer preferences for IMTA versus CCA in the producing region (British Columbia, Canada).
(If either IMTA or CCA was to be adopted for salmon farming, how strong is your preference for one method over the other?)

By making assumptions regarding the potential environmental improvements that could arise from widespread adoption of IMTA or CCA technologies in BC, the general societal benefits from their adoption were approximated. We estimated that BC residents would be willing to pay between $77.76 and $159.54 per household per year to support development and fund incentives for adoption of IMTA, and $133.28 to $173.00 per household per year to support development and fund incentives for adoption of CCA. Using the more conservative figures, this translates to a provincial willingness to pay of $143 million and $245 million per year for IMTA and CCA, respectively, based on current population estimates for the province of BC. It seems reasonable to interpret these values as indicative of the value BC residents place on improvements to marine quality in the vicinity of salmon farms. Ideally, we would like to assume these values are distinct from what BC residents would be willing to pay for IMTA products; however, there may be some conflation of these values when respondents both consume farmed salmon and value more sustainable production methods, and also value an improvement in the marine environment for its own sake.

Nonetheless, the results of this study suggest BC residents are concerned about the potential environmental impacts of salmon farming in the province, and are in favor of government policy aimed at improving the industry's sustainability. This includes funding research and development of alternative aquaculture technologies, and subsidizing companies that adopt these alternative technologies. Furthermore, while respondents are favorable towards both IMTA and CCA as new technologies, they are currently more knowledgeable and more supportive of the CCA technology than IMTA technology.

DISCUSSION

Thus far, we have concentrated on technological developments to improve environmental performance in Canadian finfish aquaculture with a focus on farmed salmon only. However, as explained earlier the introduction of IMTA involves additional "extractive" species and the impact of this new production on market supply needs to be considered as well. Taking oyster production as an example, we studied the potential increase in BC production of oysters if IMTA were adopted in the province. Our results suggest that the annual supply of oysters from IMTA operations could range widely, depending on the level of adoption and proportion of rafts dedicated to oyster production. The oyster production system presented here was a 60 raft IMTA shellfish component producing 7000 oysters per raft per year at a typical IMTA operation. Further details are available from Kitchen (2011). If as many as 45% of salmon farms adopted IMTA and one-third of associated shellfish production consisted of oysters, aggregate production at IMTA sites would increase BC oyster production by 5363 t per year. But this value rises to as much as 16,000 t per year as the share of oysters in shellfish production approaches 100%; at this level of production, the farm-gate value of oyster production at IMTA sites would be over $24 million per year, representing more than 200% of the current conventional supply of oysters from BC. Clearly, widespread adoption of IMTA in BC would have implications not just for Atlantic salmon (or other farmed finfish species), but for the marketing and sales of extractive species as well. Careful planning would be advised to ensure the development of adequate market opportunities for this increased production. See Kitchen and Knowler (2013) for a set of recommendations along these lines.

Whether such levels of adoption of IMTA in the BC salmon farming sector are likely is debatable, as noted earlier. To get a better sense of IMTA's prospects in BC we used a qualitative assessment methodology and consulted with a range of stakeholders to ascertain perceptions of the barriers to adoption of IMTA. This research is reported in Crampton (2016). We interviewed participants representing salmon aquaculture companies, industry associations, provincial and federal government regulators, and environmental non-government organizations (ENGOs). We found that participants considered uncertainty pertaining to biological and technical feasibility, fish health concerns, and the current regulatory framework, as key factors impeding IMTA adoption. Other barriers include perceived lack of profitability, preference for CCA technology, and a general lack of strong incentives. In order to address the multiple barriers that cumulatively create a strong disincentive to adopt, it was argued a "whole-of-government" approach towards IMTA would be required, perhaps including the use of market instruments to encourage socially desirable technology use.

CONCLUSION

Definitive conclusions regarding the attractiveness of IMTA as an alternative production technology for farmed salmon are elusive. For example, preference for the adoption of either IMTA or CCA for salmon farming is dependent on whether you are from a "consuming" region versus a "producing" region. Those from the producing region (BC) preferred CCA, whereas those from the consuming region (US West Coast) were more likely to prefer IMTA. Despite some indication of potential profitability, particularly for IMTA, there are barriers to the implementation of either IMTA or CCA systems. In part, this observation arises due to their status as untested and not yet having achieved full commercial scale. There are issues with management complexity, high switching costs and production risks if current salmon farmers adopt IMTA or CCA. At present, there are few if any private financial incentives for producers to adopt these technologies on a broad scale, as the biomitigation ecosystem service (and others) provided by each technology is not "rewarded". So far, feasibility studies only address profitability and ignore other economic and environmental considerations. Finally, rather than encourage technological innovation using multi-species production methods such as IMTA the new regulatory regime governing aquaculture in BC hinders adoption e.g., the Federal government is responsible for non-plant farm products (e.g. finfish, shellfish), but the province is responsible for farmed aquatic plants (e.g. kelp).

ACKNOWLEDGEMENTS

This research was funded by the Natural Sciences and Engineering Research Council of Canada (NSERC) strategic Canadian Integrated Multi-Trophic Aquaculture Network (CIMTAN), in collaboration with its partners, Fisheries and Oceans Canada, the University of New Brunswick, the New Brunswick Research and Productivity Council, Cooke Aquaculture Inc., Kyuquot SEAfoods Ltd., Marine Harvest Canada Ltd. and Grieg Seafood BC Ltd. Thanks are due to various participants in the CIMTAN program but especially Dr. Thierry Chopin. I would also like to acknowledge the graduate students who participated in, and in most cases led, the research reported here: Patrick Kitchen, Winnie Yip, Kim Irwin, Stefan Crampton and Mark Carras. It was a pleasure to work with all of them.

REFERENCES

Barrington, K., Ridler, N., Chopin, T., Robinson, S. and Robinson, B. 2010. Social aspects of the sustainability of integrated multi-trophic aquaculture. Aquaculture International, 18(2): 201-211.

Carras, M. 2017. Assessing the profitability of integrated multi-trophic aquaculture in Canada with and without a deposit feeder component. Masters of Resource Management Project No. 671, School of Resource and Environmental Management, Simon Fraser University, 134 pp.

Carras, M., Knowler, D., Pearce, C., Hamer, A., Chopin, T. and Weaire, T. (Submitted). A discounted cash-flow analysis of salmon monoculture and integrated multi-trophic aquaculture in eastern Canada. Submitted to Aquaculture Economics and Management.

Chopin, T., Robinson, S., Troell, M., Neori, A., Buschmann, A., and Fang, J. 2008. Multi-trophic integration for sustainable marine aquaculture. pp. 2463-2475. *In*: Jorgensen, S. and Fath, B. (eds). Encyclopedia of Ecology. Academic Press. Oxford.

Chopin, T. and Sawhney, M. 2009. Seaweeds and their mariculture. The encyclopedia of ocean sciences. Elsevier, Oxford, 4477-4487.

Crampton, S. 2016. Assessing the barriers and incentives to the adoption of integrated multi-trophic aquaculture in the canadian salmon aquaculture industry. Masters of Resource Management Project No. 643, School of Resource and Environmental Management, Simon Fraser University, 131 pp.

Cubillo, A.M., Ferreira, J.G., Robinson, S.M., Pearce, C.M., Corner, R.A. and Johansen, J. 2016. Role of deposit feeders in integrated multi-trophic aquaculture—a model analysis. Aquaculture, 453: 54-66.

DFO/Fisheries and Oceans Canada. 2010. Economic study of closed-containment options for the Canadian aquaculture industry. Second draft. Quebec: Fisheries and Oceans Canada.

DFO/Fisheries and Oceans Canada. 2013. Aquaculture in Canada: Integrated Multi-Trophic Aquaculture. Government of Canada, Ottawa, Canada.

Irwin, K. 2014. Valuing improvements to the environmental performance of salmon aquaculture in British Columbia: A choice modelling approach. Masters of Resource Management Project No. 593, School of Resource and Environmental Management, Simon Fraser University, 141 pp.

Kitchen, P. 2011. An economic analysis of shellfish production associated with the adoption of integrated multi-trophic aquaculture in British Columbia. Masters of Resource Management Project No. 516, School of Resource and Environmental Management, Simon Fraser University, 141 pp.

Kitchen, P. and Knowler, D. 2013. Market implications of adoption of integrated multi-trophic aquaculture: shellfish production in British Columbia. Ocean and Coasts Network (OCN) – Canada Policy Briefs, Volume 3: Round 1 (April 2013): 17-20.

Living Oceans Society. 2011. Assessing the viability of commercial-scale closed containment systems. Retrieved February 16, 2012, from Living Oceans Society: http://www.livingoceans.org/initiatives/salmon-farming/what-were-doing/assessingviability-commercial-scale-

Martinez-Espiñeira, R., Chopin, T., Robinson, S., Noce, A., Knowler, D. and Yip, W. 2016. A contingent valuation of the biomitigation benefits of integrated multi-trophic aquaculture in Canada. Aquaculture Economics and Management, 20: 1-23.

MMK Consulting Inc. 2007. Special Committee on Sustainable Aquaculture Report: Volume 1. Victoria, BC: The Legislative Assembly of British Columbia.

Ridler, N., Wowchuk, M., Robinson, B., Barrington, K., Chopin, T., Robinson, S., Page, F., Reid, G., Szemerda, M., Sewuster, J. and Boyne-Travis, S. 2007. Integrated multi-trophic aquaculture (IMTA): a potential strategic choice for farmers. Aquaculture Economics and Management, 11(1): 99-110.

Shuve, H., Caines, E., Ridler, N., Chopin, T., Reid, G., Sawhney, M., Lamontagne, J., Szemerda, M., Marvin, R., Powell, F., Robinson, S. and Boyne-Travis, S. 2009. Survey finds consumers support integrated multitrophic aquaculture. Global Aquaculture Advocate March/April 2009, 22-23.

Whitmarsh, D., Cook, E. and Black, K. 2006. "Searching for Sustainability in aquaculture: an investigation into the economic prospects for an integrated salmon-mussel production system". Marine Policy, pp. 293-298.

Wright, A. and Arianpoo, N. 2010. Technologies for viable salmon aquaculture –an examination of land-based closed containment aquaculture. http://www.saveoursalmon.ca/solutions/closed_containment: SOS Solutions Advisory Committee.

Yip, W. 2012. Assessing the willingness to pay in the pacific northwest for salmon produced by integrated multi-trophic aquaculture. Masters of Resource Management Project No. 530, School of Resource and Environmental Management, Simon Fraser University, 156 pp.

Yip, W., Knowler, D., Haider, W. and Trenholm, R. 2017. Valuing the willingness-to-pay for sustainable seafood: integrated multitrophic versus closed containment aquaculture. Canadian Journal of Agricultural Economics/Revue canadienne d'agroeconomie, 65(1): 93-117.

Chapter 11

Solutions

L.I. Bendell[1], P. Gallaugher[2], S. McKeachie[3], L. Wood[4]

[1]Professor, Biological Sciences, Simon Fraser University, Burnaby, BC
[2]Adjunct Professor, Biological Sciences, Simon Fraser University
[3]Past Chair/Director, Association of Denman Island Marine Stewards Society, Denman Island, BC
[4]Manager, Community Engagement and Research Initiatives, Faculty of Environment, Simon Fraser University

STATEMENT OF PROBLEM

Seventeen years ago DFO (Jamieson et al. 2001) recommended that regions within Baynes Sound be put aside as Marine Protected Areas (MPAs). The Sound has now been recognized as an Ecologically and Biologically Sensitive Area (EBSA) (DFO 2013) and is one of the most important ecological complexes on the Pacific west coast. DFO states, *"areas identified as EBSAs should be viewed as the most important areas where, with existing knowledge, regulators and marine users should be particularly risk averse to ensure ecosystems remain healthy and productive."* (https://open. canada.ca/data/en/dataset/d2d6057f-d7c4-45d9-9fd9-0a58370577e0). Yet, the ecological and biological significance of Baynes Sound has still not been acknowledged by either Provincial or Federal Governments whose mandate is to protect such regions.

WHAT IS SPECIAL ABOUT BAYNES SOUND?

As presented in Chapters four, five and six, the Sound is indeed one of the most important and diverse ecosystems along coastal BC. It has multiple ecological and biological roles providing key ecosystem services to a wide diversity of marine, freshwater and terrestrial flora and fauna.

Corresponding Author: L.I. Bendell

It is also home to seven of BC's red listed bird species (threatened and/or endangered) including Clark's Grebe, the Red Knot, Hudsonian Godwit, Brant's Cormorant, Cassin's Auklet, the Black-legged Kittiwake and the Common Murre (E-Fauna 2018).

WHAT ARE THE THREATS TO THE BAYNES SOUND ECOSYSTEM?

Overarching all other threats, are the catastrophic anthropogenic threats of climate change and ocean acidification that are affecting wild and farmed marine species alike, with shellfish especially vulnerable to both impacts. Given current trends, increasing ocean acidity and water temperatures casts an uncertain future on the shellfish industry.

However, and as presented in chapters four and eight, the shellfish industry currently is having a significant negative impact on the Sound with plans for even further expansion in the future (DFO 2017). Possibly the greatest impact is the continual introduction of plastic aquaculture gear such as polypropylene rope, polyethylene trays, nets, mesh, and polyvinyl chloride (PVC) pipes, into the Sound. Research into the effects of plastics on our marine environment has concluded without question, that one of the greatest threats of this century to our marine ecosystems are macro and micro plastics. Other threats to the Sound including the seaweed harvest and pollution as covered in Chapters seven and nine, are increased urbanization and the cumulative impacts that come from an increasing human population. All these threats require immediate mitigation and management.

Chapters two, three and ten, provided examples of how sensitive regions under economic pressures can be successfully managed to the benefit of all stakeholders. Indeed, on the west coast of Vancouver Island, the West Coast Aquatic Management Association (WCA) (WCA 2018) has, since 2001, successfully co-managed the economic and ecological demands on this region through a collaborative relationship among federal, provincial, Nuu-chah-nulth, and local governments. The SFU workshop of April 2014 and the subsequent WWF-Island Trust workshop of May 2018, which brought stakeholders together, were conducted with the goal of providing solutions to the management and protection of Baynes Sound as exemplified by the WCA.

Herein lies the paradox: How could one of the most ecologically and biologically sensitive regions of coastal BC continue to experience the shortcomings in management exhibited by a federal government department committed to the conservation and protection of this region?

Perhaps the reason why we have arrived at such a point of conflict is the historical context in which the shellfish industry developed within

Baynes Sound. Seventy years ago no thought would have been given to the harvest of an oyster and this approach to shellfish harvesting within the Sound remains. The industry has had longstanding, unregulated access to the Sound which has led to the continued development of this region for shellfish farming in the absence of sound ecological and biological practices.

The most recent pressure comes from the farming of geoducks. As of October 2018, 22 leases within Baynes Sound have been licensed for this purpose with more proposed (DFO 2017 and 2018a). Farming of the geoduck involves placing the developing geoducks into low inter-tidal or subtidal grow out beds where they are planted into PVC pipes covered with plastic mesh (DFO 2017). As noted in chapters four and eight, the pipes and mesh litter the beach and begin to degrade into ever smaller fragments due to UV exposure and physical processes. A review by Thornton (2009) concludes *"When the entire life cycle of PVC is considered, it is apparent that PVC is one of the most environmentally hazardous materials in production. Vinyl production, use, and disposal is responsible for generating large quantities of persistent, bioaccumulative, and toxic pollutants—and then releasing them into the global environment. Available data suggest that PVC is a significant contributor to the world's burden of persistent organic pollutants and endocrine disrupting chemicals—including dioxins and phthalates—that have accumulated universally in the environment and the bodies of the human population."*

STATEMENT OF SOLUTIONS

The most obvious solution to the conflict of Baynes Sound lies in the existing international commitments and agreements Canada has made with respect to the setting aside ten percent of our marine environment as Marine Protected Areas and in the recently adopted Oceans Plastics Charter (2018) that has committed Canada to a zero waste policy by 2020.

Canada as party to the United Nations Convention on Biological Diversity eight years ago, agreed to reduce the rate of biodiversity loss by 2020 by achieving 20 objectives known as the Aichi Targets (Lemieux et al. 2019). Of importance to Baynes Sound is Target 11 which states that *"by 2020 at least 10 percent of coastal and marine areas, especially areas of particular importance for biodiversity and ecosystem are conserved through effectively and equitably managed, ecologically representative and well-connected systems of protected areas and other effective area-based conservation measures, and integrated into the wider landscape and seascape"* (Convention on Biological Diversity 2011a).

In 2016 Canada mapped their national targets to the Aichi Biodiversity Targets with Target 1 directly in line with Target 11 of the Aichi Targets

(Convention of Biological Diversity 2011b). Further, Canada has also committed to their own Target 7, "*By 2020, all aquaculture in Canada is managed under a science-based regime that promotes the sustainable use of aquatic resources (including marine, freshwater and land based) in ways that conserve biodiversity.*" which maps directly onto Aichi Targets 4 and 7 (Convention of Biological Diversity 2011a).

Lemieux et al. 2019 has assessed Canada's attempt to meet Target 11 and concluded that in a "faltering" attempt to reach its Target 11 commitments, some Canadian jurisdictions have elected to focus more on coverage (quantity) and less on ecology integrity (quality). The authors concluded that this will have significant ramifications for long-term success of biodiversity conservation.

Baynes Sound, an area of 140 sq m^2, is roughly one tenth of the size of the proposed National Marine Conservation Area of the Southern Strait of Georgia (WWF 2018). As repeatedly documented within the chapters of the book, the Sound is one of the most ecologically and biologically sensitive regions along coastal BC. Concerns raised by Lemieux et al. (2019) are directly applicable in this case.

However, as again noted by Lemieux (2019), it is unlikely that Canada will meet its 2020 ten percent targets as the process takes years to decades to implement. Alternatively, DFO can declare Marine Refuges as "Other Effective area-based Conservation Measures (OECMs) (DFO 2018b). Five criteria that OECMs must meet are:

1) A clearly defined geographic location
2) Conservation or stock management
3) Presence of ecological components of interest
4) Long-term duration implementation
5) The ecological components of /interest are effectively conserved.

As with meeting all and more of the criteria required to be an EBSA, Baynes Sound meets all the criteria required to be established as a Marine Refuge.

Perhaps of greater urgency is the immediate action required to remove the source of and recover all derelict plastic shellfish aquaculture debris from the Sound. As with MPAs, Canada has committed itself as a signatory to the Ocean Plastics Charter (Oceans Plastic Charter 2018) and to a more resource-efficient and sustainable approach to the management of plastics. Included in the action items under Section 2, to achieve a resource-efficient lifecycle management approach to plastics in the economy are:

a. Working with industry and other levels of government, to recycle and reuse at least 55% of plastic packaging by 2030 and **recover 100% of all plastics by 2040.**

b. Increasing domestic capacity to manage plastics as a resource, **prevent their leakage into the marine environment from all sources,** and enable their collection, reuse, recycling, recovery and/or environmentally-sound disposal.

c. Encouraging the application of a whole supply chain approach to plastic production toward greater responsibility and prevent unnecessary loss, including in pre-production plastic pellets.

Canada has committed to recover 100% of all plastics by 2040 and to preventing the leakage of plastics into the marine environment from all sources. As noted in Chapter four and eight, the shellfish industry introduces both macro and micro plastics in the Sound, with six metric tonnes being recovered in fall of 2018. There must be zero tolerance for plastics entering aquatic systems which means the shellfish industry must seek alternate plastic free farming equipment and stop the flow of plastics from the source. Canada has committed to doing so, now actions must be implemented to make this happen.

IMMEDIATE ACTION REQUIRED

If there was ever an urgency to act, now, would be the time. The following are all doable and need to be promptly implemented:

1) Moratorium on all plans for the expansion of the shellfish industry within Baynes Sound.

2) Remove all plastics within the intertidal of Baynes Sound.

3) Remove all derelict shellfish aquaculture gear now present within the intertidal regions of Baynes Sound (e.g., anti-predator netting, rebar, vexar fences).

4) Restore all intertidal regions.

5) As recommended by Jamieson et al. 2001, recognize Baynes Sound as the EBSA that it is and establish it as a Marine Refuge.

6) Develop plastic-free shellfish aquaculture gear and ecologically sustainable shellfish farming practices. The First Nations approach of clam gardens as discussed by Bendell (2015) would be appropriate here.

7) With a team of academic and government scientists, identify regions where ecologically sustainable shellfish farming can occur within the Marine Refuge of Baynes Sound.

Solutions are at hand. An action plan for the implementation of these solutions needs to be put in place immediately.

REFERENCES

Bendell, L.I. 2015. Favored use of anti-predator netting (APN) applied for the farming of clams leads to little benefits to industry while increasing nearshore impacts and plastics pollution. View Point. Marine Pollution Bulletin. 91: 22-28.

Convention on Biological Diversity. 2011a. https://www.cbd.int/. Accessed November 24, 2018.

Convention of Biological Diversity. 2011b. https://www.cbd.int/countries/targets/?country=ca. Accessed November 24, 2018.

DFO. 2013. Evaluation of proposed ecologically and biologically significant areas in marine waters of British Columbia. DFO Canadian Scientific Advisory Secretariat Science Advisory Report. 2012/075.

DFO. 2017. Integrated Geoduck Management Framework, Pacific Region 2017. 19 pp. http://www.pac.dfo-mpo.gc.ca/aquaculture/management-gestion/geoduck-panope/index-eng.html Accessed November 25, 2018.

DFO. 2018a. Current valid British Columbia Shellfish Licenses. Aquaculture Management. https://open.canada.ca/data/en/dataset/522d1b67-30d8-4a34-9b62-5da99b1035e6. Accessed October 28, 2018.

DFO. 2018b. Other Effective Area-Based Conservation Measures: Creating Marine Refuges in Canada. http://www.dfo-mpo.gc.ca/oceans/documents/oeabcm-amcepz/marineshelters-refugesmarins-eng.pdf Accessed November 24, 2018.

E-Fauna BC. 2018. Electronic Atlas of the Wildlife of BC. http://ibis.geog.ubc.ca/biodiversity/efauna/ Accessed November 24, 2018.

Jamieson, G.S., Chew, L., Gillespie, G., Robinson, A., Bendell-Young, L., Heath, W., Bravender, B., Nishimura, D. and Doucette, P. 2001. Phase 0 review of the environmental impacts of inter-tidal shellfish aquaculture in Baynes Sound. Fisheries and Oceans Canada, Canadian Science Advisory Secretariat, Ottawa, Ontario. 103 p.

Lemieux, C.J., Gray, P.A., Devillers, R., Wright, P., Dearden, P., Halpenny, E.A., Groulx, M., Beechey, T.J. and Beazley, K. 2019. How the race to achieve Aichi Target 11 could jeopardize the effective conservation of biodiversity in Canada and beyond. Marine Policy 99: 312-323.

Oceans Plastics Charter. 2018. G7, Charlevoix, Quebec Canada. https://g7.gc.ca/wp-content/uploads/2018/06/OceanPlasticsCharter.pdf Accessed November 24, 2018.

Thornton, J. 2009. Environmental Impacts of Polyvinyl Chloride (PVC) Building Material. A briefing paper for the Health Building Network. Columbia Earth Institute, Columbia University. 79 pp.

WCA. 2018. http://westcoastaquatic.ca/about-us/ Accessed November 24, 2018

WWF. 2018. http://www.wwf.ca/conservation/oceans/marine_protected_areas/. Accessed November 24, 2018.

Index

About the Editors

Dr. Leah Bendell is a Professor in the Department of Biological Sciences, Faculty of Science. She has for the past 30 years studied how anthropogenic impacts alter ecosystem structure and function and the consequences of such impacts on ecosystem and human health. Her research has taken her to the Albertan Tar Sands, the mines of Indonesia, the freshwater lakes and wetlands of Ontario and the intertidal regions of coastal British Columbia. She teaches Ecotoxicology and Biology and tries to instill in her students a sense of urgency as to why we must all work together to mend our planet.

Dr. Patricia Gallaugher is an adjunct professor in the Department of Biological Sciences at Simon Fraser University. Founder and former director of the Centre for Coastal Studies and Management and the Speaking for the Salmon programs in the Faculty of Environment; over the past 25 years she has conducted community engagement programs and research focussed on seeking solutions for sustainability and ocean conservation coast-wide in British Columbia. She is the recipient of the Roderick Haig Brown Award for Conservation and the Murray Newman Award for Excellence in Aquatic Conservation and Research.

Shelley McKeachie graduated from the Simon Fraser University Professional Development Program in 1977 followed by a career as an educator for 30 years. She became a Project Wild facilitator and taught both teachers and students about the importance of the natural world. Shelley is a founding member, past chair and director of the Association for Denman Island Marine Stewards (ADIMS), working for 18 years to raise awareness of the importance of responsible management and protection of the marine environment.

Laurie Wood is the Manager of Community Engagement and Research Initiatives for the Faculty of Environment at Simon Fraser University. For over two decades, she has organized multi-sectoral dialogues with an aim to link science and local knowledge to inform conservation management and practice. This has involved fostering and managing relationships among a diverse group of partners including students, faculty, and representatives from First Nations, government, industry, and community groups.

Color Plate Section

Chapter 1

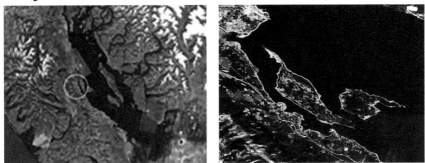

Figure 1 Baynes Sound and Lambert Channel.

Chapter 3

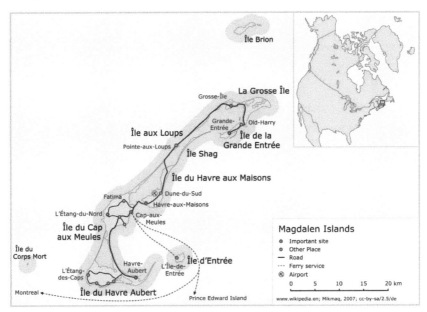

Figure 1 Magdalen Islands. Image from Wikepedia. Accessed November 2018.

Chapter 4

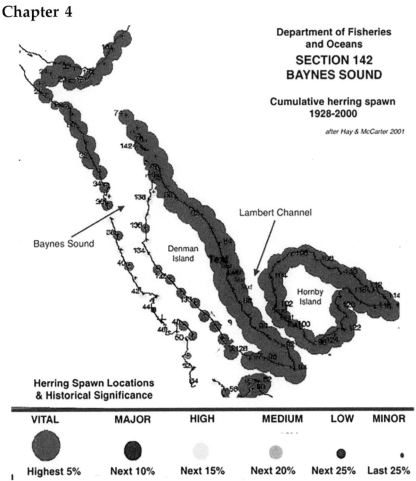

Department of Fisheries and Oceans
SECTION 142
BAYNES SOUND

Cumulative herring spawn 1928-2000

after Hay & McCarter 2001

Lambert Channel

Baynes Sound

Denman Island

Hornby Island

Herring Spawn Locations & Historical Significance

VITAL	MAJOR	HIGH	MEDIUM	LOW	MINOR
Highest 5%	Next 10%	Next 15%	Next 20%	Next 25%	Last 25%

Figure 1   Herring spawn locations within Baynes Sound, Lambert Channel.

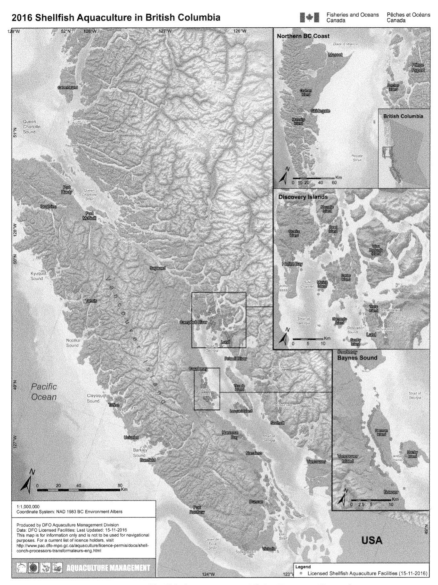

Figure 2 Shellfish tenures within Baynes Sound.

Chapter 8

Figure 1   Example of tonnage of debris collected in the fall 2017 beach clean-up. Ninety percent of debris is due to the shellfish industry.

(a)   **(b)**

Figure 2a&b   (a) Installation of geoduck piping and
(b) Debris left behind post installation.

Chapter 9

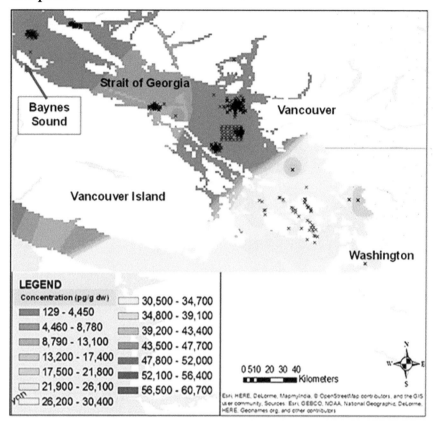

Figure 2 Spatial variation of ΣPCB average concentration (values in pg/g dry weight) and distribution of hot spots in sediments in south BC and Washington estimated using inverse distance weighted (IDW) interpolation method in Arc-GIS. Samples sites/locations are indicated with black x. Based on and adapted from Pearce and Gobas (2018), Pearce (2018).

Chapter 10

Figure 1   Integrated multi-trophic aquaculture (IMTA).

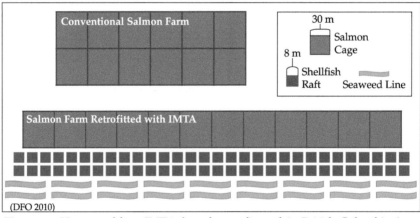

Figure 2   How would an IMTA farm be configured in British Columbia (versus New Brunswick) (Kitchen 2011)?